A Concise Glossary of

CONTEMPORARY LITERARY THEORY

SECOND EDITION

A Concise Glossary of CONTEMPORARY LITERARY THEORY

SECOND EDITION

JEREMY HAWTHORN

Professor of Modern British Literature,
University of Trondheim, Norway

Edward Arnold
A member of the Hodder Headline Group
LONDON NEW YORK MELBOURNE AUCKLAND

This new edition is again for Richard, Joanna, Dinah and Nancy, who continue to put us up and up with us.

First published in Great Britain 1994

British Library Cataloguing in Publication Data

Available on request.

ISBN 0 340 601876

Typeset in Times by the author.
Printed and bound in Great Britain for Edward Arnold, a division of Hodder Headline PLC,
338 Euston Road, London NW1 3BH by Biddles Ltd, Guildford and King's Lynn

Introduction to the Second Edition

Vincent M. Colapietro has a good opening sentence in his *Glossary of Semiotics* (Paragon House, 1993): 'In composing this glossary, I learned what I should have already known'. Me too. I can add that in composing a second edition I now have to reveal what I still did not know when I completed the first edition. I feel embarrassed by a number of my areas of ignorance; for example, my extremely cursory account of cultural materialism – surely one of the most interesting and productive of theoretical groupings to have emerged during the last decade or so. The same is true of my earlier brief entry on hermeneutics. I have to admit that the term *thick description* was unknown to me in 1991, that *obstination* was something that I would probably have associated with British government policy on education rather than with Free Indirect Discourse, and that I would not even have been able to hazard a guess as to the meaning of the Uncle Charles Principle.

This is beginning to sound a bit like the academic game of humiliation, amusingly described by David Lodge in his novel *Changing Places*, a game in which the person who can admit to not having read the most important work of literature wins the game. However, although I do feel embarrassed about some of my gaps of knowledge, I don't feel absolutely humiliated. Literary and critical theory today takes its concepts, methods and terms from such a wide range of disciplines and areas of study that to keep abreast of all new developments would involve the sort of study that would probably make it hard to read any literature at all – a situation which I for one would find very unsatisfactory.

The question that I have been asked most often by readers of the glossary is, 'What is the reason for the explosion in literary theory that has taken place in Britain and the United States during the past two decades?'. Up to the 1950s, twentieth-century Britain was known as the home of pragmatism and empiricism, the venue of applied studies in the Humanities and the Social Sciences, a place set apart from a Continent (which naturally excluded Britain!) in which people had nothing better to do than to ponder over all manner of abstruse and recondite theoretical questions. What has happened to change this situation?

One of the conclusions that influential Russian psychologists such as Lev Vygotsky and Alexander Luria reached following their investigations into human maturation processes was that abstraction follows when individuals meet problems which cannot be solved in more pragmatic and empirical ways. It seems to me highly likely that what is true of individuals is also true of cultures and nations – and academic disciplines. In his classic essay 'Components of the

National Culture' (1969), Perry Anderson associates the traditional antipathy displayed towards abstract theory in British intellectual life with the success of British imperialism. A country running a successful world empire was occupied with the practical problems associated with its upkeep more than with those abstract and theoretical concerns which lacked immediate practical implications. There is certainly no doubt that the turn to theory coincides with the intensification of a whole range of social, political and economic crises in Britain and, in very different forms, in the United States. It is also associated with a more delimited set of crises – those experienced by the political Left during the past two decades. Much (but by no means all) recent theory has a pinkish tinge, and it is not hard to understand that the setbacks and disappointments experienced by the political Left during the past two decades in Britain and elsewhere have turned the attention of many to more fundamental and abstract questions.

Other extra-literary factors may also be relevant. The explosion of interest in narrative goes along with a more general concern with 'who is talking to (or at) whom?' in modern societies, a concern that is certainly related to the work of analysts of the mass media. And it would be impossible to explain the growth of feminist theory without a consideration of the social and cultural forces which lay behind the rebirth of the Women's Movement.

But it is also the case that shifts in academic subjects can occur for internal as well as for external reasons, and that when they take place the upheaval associated with them often generates theoretical activity. Jonathan Culler is not the only literary critic and theorist who has suggested that by the 1970s much literary criticism seemed stuck in a repetitive process of generating more and more interpretations of a limited set of canonical texts.

Antagonism to theory is not rare amongst students today, and this should give us pause for thought. How is it that that turn to theory which so many students and teachers of my generation found liberating in the 1960s and 1970s is now seen as a chore and an irrelevance by some students? There is more than one answer to this question, but I think that one major reason is that all too often students are prevented from adopting a *problem-centred* approach to the study of theory. Theory is thus cut off from that which it is supposed to be the theory *of*; it becomes another set of facts (schools, critics, terms) which have to be learned before one can get back to what one really wants to do – read and study literature.

A good portion of what is new in this second edition is there thanks to readers of the first edition who sent me their constructive and diplomatic comments and suggestions. Many other valuable corrections to the first edition have been sent to me by its German translator, Dr Waltraud Kolb. I would like to thank all who have made suggestions or pointed out errors. I hope that the second edition bears evidence of their concern.

Jeremy Hawthorn
University of Trondheim, Norway
January 1994

Using the Glossary

1 Conventions of presentation

I have attempted to group related terms in common entries so as to avoid repetition and so as to permit entries that have a certain completeness. Thus rather than having separate entries for *plot*, *story*, *fabula* and *sjužet*, I have a single entry entitled STORY AND PLOT. The negative result of such a policy, inevitably, is that those searching for the meaning of terms must frequently tolerate having to find their way to the substantive entry via cross-entries. I hope that the gains of this procedure will be deemed to outweigh the losses.

The use of small capitals (e.g. DECONSTRUCTION) betokens the existence of an entry on the term so presented. Sometimes the actual entry may be under a cognate term; thus DECONSTRUCT refers the reader to the entry for the term DECONSTRUCTION. I have normally limited this use of small capitals to the *first* mention only of the term in question within each entry. To avoid confusion the use of small capitals generally directs the reader to substantive (rather than cross-reference) entries, although I have made exceptions to this rule in the case of particularly important terms or connections.

Bracketed references with page numbers refer the reader to the bibliography at the end of the book, which provides full publication details of works quoted from and referred to.

Many of the terms with which I deal are of foreign origin, and often it has had to be a matter of personal judgement as to whether they are sufficiently assimilated to be written without the use of the italics that indicate a loan word. I have based my decision here upon the extent to which the words in question can be said to be in frequent use amongst theorists: thus I give *épistémè* and *méconnaissance* italics, but not écriture or flâneur. In transliterating Russian names and terms I have adopted the more traditional versions (e.g. Eichenbaum rather than Èjxenbaum), unless I am referring to a published source. Thus both of these variants appear in the bibliography and in textual references.

American readers should note that British spelling is used throughout except where I quote from a source in American English. Thus there is no entry for CENTER, but there is one for CENTRE. In addition, the bibliography generally gives the place of British publication where one exists.

This glossary is also available in hardback in an extended, reference edition, which includes a number of more general entries, and expands certain others.

2 Criteria for selection of terms

My aim in the pages that follow has been to provide the reader with sufficient information to enable him or her to make sense of those of the more common specialist terms used by recent literary critics or theorists which cannot be found in more general dictionaries or glossaries of literary terms. By 'recent' I mean, generally, from about 1970, although I have occasionally allowed myself to be inconsistent on this point. Terms that have entered Anglo-American theoretical discussion via recent English translations of older works (by Bakhtin and Ingarden, for example) are also included.

3 Schools and approaches

As an additional aid to those using this glossary I append here a list of most of the terms included in it grouped according to their intellectual associations or origins. Where terms are asterisked this is to suggest either that they represent concepts central to the grouping in question, or that the entry thus indicated contains information about the school or approach. Those wanting a fuller account of such groupings should consult Ann Jefferson and David Robey (eds), *Modern Literary Theory* (2nd edn, Batsford, 1986), or Raman Selden's and Peter Widdowson's *A Reader's Guide to Contemporary Literary Theory* (rev. 3rd edn, Harvester Wheatsheaf, 1993).

Anthropology and Cultural Studies
*binary / binarism; **bricoleur*; *culture; formulaic literature; *myth; *structures of feeling; *thick description / thin description

Bakhtin group
assimilation; *carnival; *centrifugal / centripetal; character zone; *chronotope; contiguity; *dialogic; discourse; dominant; enthymeme; exotopy; *heteroglossia; horizon; hybrid; orchestration; *polyphonic; refraction; reification; semantic position; *skaz*; transgredient; utterance

Deconstruction
agon; aporia; *arche-writing; author; *centre; Copernican revolution; *deconstruction; desire; *différance; dissemination; echolalia; *écriture; ephebe; erasure; *grammatology; hinge; *logocentrism; *logos; *ludism; *mise-en-abyme*; phallogocentrism; *post-structuralism; *presence; *radical alterity; reference / referent; revisionism; **s'entendre parler*; site; subject and subjectivity; textualist; *transcendental pretence / signified / subject

Discourse analysis
*archaeology of knowledge; closure; *discourse; *dispositif*; **épistémè*; exteriority; genotext and phenotext; New Historicism and cultural materialism; *signifying practice; slippage; speech; suture; text and work; topic; *utterance

Feminism
androcentric; *androgyny; *desire; *difference; *écriture féminine; *female affiliation complex; *feminism; *gaze; *gender; gynocratic; *gynocritics; *immasculation; logic of the same; *marginality; *muted; other; *patriarchy; *phallocentrism; pleasure; pornoglossia; quest narrative; reading position; realism; *sexism; *stereotype; subject and subjectivity

Linguistics
actualization; *arbitrary; aspect; *competence and performance; cratylism; *diachronic and synchronic; diacritical; difference; *discourse; *displacement; *functions of language; idiolect; *langue and parole; linguistic paradigm; markedness; metalanguage; punctuation; register; Sapir-Whorf hypothesis; shifter; *sign; sociolect; speech; *speech act theory; *syntagmatic and paradigmatic; text and work

Marxism
absence; *alienation; aura; *base and superstructure; Copernican revolution; *dialectics; *epistemological break; *fetishism; flâneur; *hegemony; homology; ideologeme; *ideology; incorporation; instance; *intellectuals; *interpellation; legitimation; *Marxist literary theory and criticism; *materialism; moment; montage; myth; popular; problematic; *realism; *reification; slippage; structure in dominance; subject and subjectivity

Media Studies
*agenda setting; digital and analogic communication; *gatekeeping; hot and cool media; *technological determinism; *uses and gratifications

Narratology
achronicity / achrony; act / actor; actualization; anachrony; analepsis; architext; aspect; cancelled character; *connotation and denotation; crisis; defamiliarization; deferred / postponed significance; deixis; deviation; *diegesis and mimesis; digital and analogic communication; *discourse; distance; duration; ellipsis; enunciation; event; figure; figure and ground; flicker; flip-flop; frame; *Free Indirect Discourse; frequency; function; hinge; homology; homonymy; interior dialogue; interpolation; *intertextuality; intrusive narrator; isochrony; *linguistic paradigm; metalepsis; *mise-en-abyme*; mode; montage; mood; *narratee; narration; *narrative; *narrative situation; *narratology; obstination; order; paralipsis; *perspective and voice; power; privilege; prolepsis; repetition; script; semantic axis; *skaz*; short-circuit; slow-down; *story and plot; suspense; suture; syllepsis; synonymous characters; *text and work; theme and thematics; Uncle Charles Principle

New Historicism and Cultural Materialism
circulation; *emplotment; energy; exchange; *New Historicism and cultural materialism; resonance

Phenomenology and Geneva School
epoché; *phenomenology; polyphonic

Pragmatics
*discourse; double-bind; *politeness; *pragmatics; *speech act theory

Prague School / Linguistic Circle
actualization; *concretization; *deformation; *dominant; *literariness; norm; *Prague School

Psychology and Psychoanalysis
body; *condensation and displacement; contiguity; crosstalk; desire; double-bind; *fetishism; figure and ground; *fort / da; gaze; hommelette; imaginary / symbolic / real; *jouissance; *linguistic paradigm; *méconnaissance*; *mirror stage; *nachträglichkeit*; name-of-the-Father; *object-relations theory / criticism; *objet a / objet A;* other; overdetermination; phallocentrism; pleasure; primary process; repression; revisionism; solution from above / below; syntagmatic and paradigmatic; *topographical model of the mind; *transference; *unconscious

Reader-response criticism
code; ecological validity; hermeneutics; *interpret[at]ive communities; inter-subjectivity; jouissance; *meaning and significance; *open and closed texts; oppositional reading; *readers and reading; *reading community; *reading position; *reception theory; *self-consuming artifact;

Russian Formalism
*defamiliarization; *deformation; deviation; *dominant; fantastic; figure and ground; *function; *functions of language; *literariness; *Russian Formalism

Semiotics and Information Theory
autopoiesis; *binary / binarism; *code; *digital and analogic communication; echolalia; *epoché*; myth; *redundancy; sememe; *semiology / semiotics; *Shannon & Weaver model of communication; *sign; signifying practice

Structuralism and post-structuralism
*arbitrariness; *author; **bricoleur*; convention; deviation; *diachronic and synchronic; diacritical; *difference; digital and analogic communication; *écriture; *formulaic literature; *function; *functions of language; homology; *langue and parole; *linguistic paradigm; *post-structuralism; *sign; speech; structure in dominance; syntagmatic and paradigmatic; *structuralism; transgressive strategy

Style and stylistics
closure; commutation test; *connotation and denotation; deviation; kernel word or sentence; punctuation; *style and stylistics; terrorism; text and work

4 Other useful glossaries and reference books

M. H. Abrams's *A Glossary of Literary Terms* appears regularly in revised editions and has intelligent entries that are both accessible and critically sophisticated. The 6th edition was published in 1993 by Harcourt Brace Jovanovich. J. A. Cuddon's *A Dictionary of Literary Terms* has also appeared in a new edition published by Blackwell (1990), and in paperback by Penguin (1992). It contains a much larger number of entries than the Abrams glossary, but these are, inevitably, rather less detailed than those in Abrams. *The Concise Oxford Dictionary of Literary Terms* by Chris Baldick (Oxford University Press, 1990) places its emphasis on the succinct explanation of 'those thousand terms that are most likely to cause the student . . . some doubt or bafflement'.

Bernard Dupriez's *A Dictionary of Literary Devices* (Harvester Wheatsheaf, 1991) has been translated from the French and adapted by Albert W. Halsall. It offers perceptive and often amusing definitions of a range of terms taken from Linguistics, Prosody, Rhetoric, and Philology. Richard A. Lanham's *A Handlist of Rhetorical Terms* has now appeared in a second edition (University of California Press, 1991). The terms are very often quite technical, but the explanations are very clear and well-illustrated.

Also to be recommended is *The Longman Dictionary of Poetic Terms* by Jack Myers and Michael Simms (1989). *Critical Terms for Literary Study*, edited by Frank Lentricchia and Thomas McLaughlin (University of Chicago Press, 1990) contains essay-length entries on 22 central terms in use in Literary Studies, including *interpretation*, *figurative language*, *author*, *canon*, and *discourse*. The standard of the essays (all by different contributors) is high.

It will be seen that with all of these works one chooses between breadth and depth: the more entries, the shorter and less detailed they tend to be.

Of the more specialist sources, the following are worth noting.

Bakhtin group

M. M. Bakhtin, *The Dialogic Imagination*, Michael Holquist (ed.), (University of Texas Press, 1981) contains a useful 11-page glossary of some of the many coinages for which Bakhtin is responsible. Tzvetan Todorov's *Mikhail Bakhtin: The Dialogical Principle* (University of Minnesota Press, 1984) contains many illuminating discussions of Bakhtin's more idiosyncratic terms.

Cultural Studies

Raymond Williams's *Keywords* (Fontana, 1976) describes itself in its sub-title as 'a Vocabulary of Culture and Society', and makes fascinating reading. The concentration upon the origins and historical changes in meaning of key terms is especially illuminating. The edition presently available (1988) was revised by Williams not long before his death.

Feminism
There is not, to my knowledge, a good glossary of feminist terms which is specifically aimed at students of literature. Maggie Humm's *The Dictionary of Feminist Theory* (Harvester, 1989) covers a number of terms, but some of the entries are rather oversimplified. *Feminism and Psychoanalysis: A Critical Dictionary* (Blackwell, 1992) is highly recommended. It is edited by Elizabeth Wright, and contains a large number of well-informed and clearly written entries from a range of contributors.

Linguistics and Stylistics
Katie Wales, *A Dictionary of Stylistics* (Longman, 1989) contains useful explanations of many of the specialist terms from Linguistics and Stylistics which are likely to be of interest to students of literature and literary theory. R. L. Trask, *A Dictionary of Grammatical Terms in Linguistics* (Routledge, 1993) is likely to become the standard reference work in its field.

Narratology
Gerald Prince, *A Dictionary of Narratology* (Scolar Press, 1988) is highly recommended. Most of the entries are relatively short, but they are clear and detailed, and the dictionary is very comprehensive.

Media Studies
James Watson and Anne Hill, *A Dictionary of Communication and Media Studies* (3rd edn, Edward Arnold, 1993), gives concise definitions of most of the specialist terms from this area which are likely to be of interest to students of literature and literary theory.

Russian Formalism and the Prague School
A very useful source here is L. M. O'Toole and Ann Shukman, 'A Contextual Glossary of Formalist Terminology', which is to be found in the journal *Russian Poetics in Translation*, vol. 4, 1977, pp. 13–48. The entries consist of brief quotations from the key texts, grouped under central terms and concepts.

Semiotics
Vincent M. Colapietro, *Glossary of Semiotics* (Paragon House, 1993) is a handy guide, although the entries are generally short and contain little bibliographic detail.

Structuralism and Post-structuralism
Although not a dictionary or a glossary, Richard Harland's *Superstructuralism: The Philosophy of Structuralism and Post-structuralism* (Methuen, 1987) contains useful and intelligent discussions of many of the relevant central terms.

Absence An interest on the part of READERS and critics in what is not to be found in a literary WORK as against what is, did not suddenly emerge in the present century: a concern to note what is lacking in one or more of an AUTHOR'S works seems to be a natural component of literary-critical discussion. But recent theorists have drawn more particular attention to this issue, and since the publication of Pierre Macherey's *Pour une Théorie de la Production Littéraire* (1966; translated as *A Theory of Literary Production*, 1978) such absences have been accorded more overt theoretical attention. According to Macherey the book is not self-sufficient but is necessarily accompanied by a certain absence without which it would not exist, and he draws our attention to the fact that Freud relegated the absence of certain words to the UNCONSCIOUS. Perhaps not surprisingly, the more a critic or theorist sees the author as in less than complete conscious control over his or her creation, the more likely it is that absences from the work will be seen to be significant. At the time of writing this work Macherey was a disciple of the French MARXIST philosopher Louis Althusser, and Althusser had argued that novels could allow us to see (but not know) the IDEOLOGY from which they were born and in which they bathed, from the inside (1971, 204). In like manner Macherey, and others following him, gave the concept of absence a specifically ideological importance. As it is seen by such theorists to be typical of ideologies that they are unable to confront their own conditions of existence, any ideology imposes blank spots and absences upon those in its grip. Thus by a process of logical retracing of steps it should be possible to read off the ideological underpinnings of a work by isolating its significant absences. From this perspective a work's absences are as significant as was the dog that did not bark to Sherlock Holmes.

An absence can, according to such theorists, be *determinate*. Thus using both Althusser and Macherey, Graham Holderness (1982, 12) has argued that the determinate absence of D. H. Lawrence's novel *Sons and Lovers* is the bourgeois class; this is the missing element which forms and controls the novel, and which has to be perceived in order fully to understand it. There is no direct engagement with the bourgeoisie in the novel, but the importance of that class to Lawrence and in his society ensures that even its exclusion from the novel is determining. This is an absence on the level of content, but absences may also occur with regard to formal and technical matters: in the final section of

James Joyce's *Ulysses* or in the poetry of e e cummings we notice the absence of many conventionally expected PUNCTUATION marks.

The Freudian concept of *lack*, associated most with the castration complex and the female's definition in terms of her lack of a penis and her attendant penis-envy, has not surprisingly attracted a fair amount of FEMINIST criticism. Some feminist writers have suggested that for the female author there are writerly parallels to the Oedipus complex: see the entry for FEMALE AFFILIATION COMPLEX.

See also OPPOSITIONAL READING ('hermeneutics of suspicion').

Achronicity / achrony Achronicity is a state in which temporal relationships cannot be established; applied to a NARRATIVE it implies the impossibility of establishing an accurate chronology of events. An achrony is an EVENT in a narrative which cannot be located on a precise time scale, cannot be temporally related to other events in the narrative.

Thus the sentence, 'John and Albert both had unhappy love affairs with the same woman' informs us that two unhappy love affairs took place, but nothing about which came first (or whether both were simultaneous).

Not to be confused with ANACHRONY.

See also ISOCHRONY.

Act / actor Bernard Dupriez provides a neat diagram of the ways in which Etienne Souriau, Vladimir Propp, and A. J. Greimas – three important narrative theorists – see the varied 'actantial paradigms':

SOURIAU	PROPP	GREIMAS	EXAMPLES
directed power	hero	subject	philosopher
object of desire	princess	(desired) object	the world
desired obtainer	dispatcher	dispatcher / donor	God
antagonist		receiver / beneficiary	humanity
arbiter / referee	aggressor / false hero	opposer	matter
helper	helper	helper	the Mind

(After Dupriez 1991, 16)

We can define an actor as an agent in a NARRATIVE that performs actions, while for an agent in a narrative to act is for that agent to cause or experience an EVENT (Bal 1985, 5). An actor does not have to be an individual, or even human. Steven Cohan and Linda M. Shires (1988) suggest a main division of actors into *subject* and *object* rôles, depending upon whether they act or are acted upon. They further suggest four additional categories of actor who have an indirect relation to events: *sender*, *receiver*, *opponent* and *helper* (1988, 69). These categories clearly owe something to the influence of the SHANNON & WEAVER MODEL OF COMMUNICATION. The French terms *destinateur* and *destinataire* (or dispatcher and donor) are, following the narrative theorist A. J. Greimas, often used instead of *sender* and *receiver*.

A number of theorists use the term *actant* rather than actor, although Prince defines an actant as a rôle rather than as an agent (1988, 1). Thus as Genette (following Spitzer) points out, any narrative in autobiographical form divides the subject of the autobiography into two actants: the narrating I and the narrated I (1980, 252).

Actualization Many contemporary linguistic and literary theories distinguish between underlying abstract SYSTEMS and the particular implementations, manifestations, or *actualizations* which these enable or from which they are generated. Thus PAROLE can be seen as an actualization of LANGUE, PERFORMANCE of COMPETENCE, the particular literary READING of a general literary competence, a given folk tale of a set of possibilities contained in the set of FUNCTIONS, and so on. In a general sense PRAGMATICS can be defined as the study of actualizations.

What generally characterizes systems and their actualizations in such subjects as Linguistics and Literary Criticism is that the implementations are much richer than the formalized system: in other words, that the real, actual system which researchers assume lies behind the actualizations is more complex and extended and has greater generative force than the formalized systems which researchers have been able to construct. No grammarian has been able to construct a grammar that can unfailingly distinguish between grammatical and non-grammatical utterances with the degree of accuracy of a native speaker of the language in question. Thus study of actualizations normally feeds back to system-construction: we refine our systems partly through abstract work, but also by adapting them to the evidence acquired from pragmatic investigations.

In the model constructed by Claude Bremond (1966), a NARRATIVE consists of functions each of which opens for two possible alternatives: actualization and non-actualization. Thus each function introduces two possible directions for the STORY to take.

In some usages *actualization* is interchangeable with CONCRETIZATION: see this entry for a more detailed account of what is meant by a reader's concretization of a literary work. Some writers have also used *actualization* as a synonym for FOREGROUNDING because of its similarity to the original Czech form of the latter term, but this usage is not common.

Addressee / addresser See FUNCTIONS OF LANGUAGE

Aesthetic norm See NORM

Agenda setting From committee procedure: control of what goes on the agenda of a meeting can guarantee that certain things are not even considered.

Agon From the Greek meaning a contest: the part of a classical Greek drama in which the chorus splits in two to support two protagonists engaged in verbal debate. The term is invoked by Harold Bloom in his collection of essays of the same name (1982), especially in the chapter 'Agon: Revisionism and Critical Personality' in which he develops his ideas of misreading and misprision.

For Bloom, REVISIONISM unfolds itself only in fighting, it is a spirit which portrays itself as agonistic, and he relates this to 'the American religion of competitiveness, which is at once our glory and (doubtless) our inevitable sorrow' (1982, viii). Revisionism, then, for Bloom, carries on a tradition of struggle that goes back to the beginning of our CULTURE and our literature.

Alienation In its more specifically Marxist sense alienation refers to an experience believed by Marx to be the result of the development of the capitalist mode of production, wherein the worker was related to the product of his or her labour 'as to an alien object', and that 'the more the worker spends himself, the more powerful becomes the alien world of objects which he creates over and against himself, the poorer he himself – his inner world – becomes, the less belongs to him as his own. . . . Whatever the product of his labour, he is not' (Marx 1970b, 108).

In discussion of MODERNIST literature the term has come to be used in a rather more general sense to characterize the sense of 'non-belonging', exclusion and loneliness seen to be typical of the modernist vision.

Some translations of the work of Mikhail Bakhtin use *alien* and cognate terms to represent Bakhtin's view of language as stamped with and soiled by the ownership marks of other people. See the entry for DIALOGIC.

Alterity See EXOTOPY; RADICAL ALTERITY

Anachrony Also, following Bal (1985), *chronological deviation.* Any lack of fit between the order in which events are presented in the PLOT or SJUŽET and that in which they are reported in the STORY or FABULA is termed an anachrony. Both ANALEPSIS and PROLEPSIS are examples of anachrony.

Bal isolates two sorts of anachrony: *punctual anachrony*, when only one instant from the past or future is evoked, and *durative anachrony*, when a longer span of time or a more general situation is evoked.

See also DURATION.

Analepsis Also, following Prince (1988), *flashback*, *retrospection*, *retroversion*, *cutback* or *switchback*. An analepsis involves 'any evocation after the fact of an event that took place earlier than the point in the story where we are at any given moment' (Genette 1980, 40).

As with PROLEPSIS the inclusion of evoked as well as narrated events extends the reach of analepsis rather beyond that traditionally accorded to a term such as flashback in pre-STRUCTURALIST days.

An *internal analepsis* does not go back beyond the chronological point at which the STORY started, while an *external analepsis* does. A *completing analepsis*, according to Genette, fills in a gap or ELLIPSIS left earlier on in the NARRATIVE, while a *repeating analepsis* or *recall* repeats that which has already been narrated. Analepses can be measured according to their *extent* (how long a period of time they cover), and *reach* (how far back in time they go) (Genette 1980, 48; Prince 1988, 5).

See also DIEGESIS AND MIMESIS (heterodiegesis and homodiegesis).

Analogic communication See DIGITAL AND ANALOGIC COMMUNICATION

Androcentric Literally: centred on the male. The term has been coined by FEMINIST theorists wishing to describe a habit of mind and set of attitudes which are based upon a male perspective and which ignore female experience and interests. The opposite of androcentric is *gynocentric*: centred on the female. Gynocentricity is advocated by some feminists as a counter-balance to androcentricity, and in the field of literature requires writers and READERS to attempt to ground themselves on female experience and to view the world from a female perspective.

Androcratic See GYNOCRATIC

Androgyny Technically, the union of both sexes in one individual. The OED gives this as a biological term and equates it with hermaphrodism, but in recent FEMINIST writing the term is used to refer to CULTURALLY acquired characteristics rather than to biologically determined ones.

The writer who probably contributed most to this shift of emphasis was Virginia Woolf. Towards the end of her long essay, *A Room of One's Own* (1929), she reports the train of thought inspired in her by looking out of her window on what she claimed was a particular day (October 26, 1928) and seeing a taxi-cab stopping for a girl and young man, picking them both up, and driving off.

> [T]he sight of the two people getting into the taxi and the satisfaction it gave me made me also ask whether there are two sexes in the mind corresponding to the two sexes in the body, and whether they also require to be united in order to get complete satisfaction and happiness? . . . The normal and comfortable state of being is that when the two live in harmony together, spiritually co-operating. If one is a man, still the woman part of the brain must have effect; and a woman

> also must have intercourse with the man in her. Coleridge perhaps meant this when he said that a great mind is androgynous. (Woolf 1929, 147–8)

Those who have sought to use and develop Woolf's suggestion have generally paid less attention to the GENDER of writers and more to the gender of, as it were, their productions – that is to say, to the attitudes, IDEOLOGIES and assumptions encoded in their writing.

Not all recent feminists have been happy with such an approach. While some have seen its acceptance and development as a way to break down SEXIST divisions between genders, others – especially radical feminists (see the entry for FEMINISM) – have reacted to it far more critically. Mary Daly, for example, commenting upon the word *androgyny*, claims that

> Experience proved that this word, which we now recognize as expressing pseudowholeness in its combination of distorted gender descriptions, failed and betrayed our thought. . . . When we heard the word echoed back by those who misinterpreted our thought we realized that combining the 'halves' offered to consciousness by patriarchal language usually results in portraying something more like a hole than a whole. (Daly 1979, 387)

And K. K. Ruthven has quoted Adrienne Rich's not unrelated objection, that 'the very structure of the word replicates the sexual dichotomy and the priority of *andros* (male) over *gyne* (female)' (Ruthven 1984, 106; quoting from Rich 1976, 30). In response to such criticisms, Sandra Gilbert and Susan Gubar have suggested the alternative term *gyandry*, but this coinage has not succeeded in displacing *androgyny*.

Anisochrony See ISOCHRONY

Anticipation See PROLEPSIS

Anxiety of influence See REVISIONISM

Apophrades See REVISIONISM

Aporia From the Greek for an apparently irresolvable logical difficulty, this term was traditionally used to describe statements by CHARACTERS in just that state – normally in soliloquy. (Hamlet's 'To be or not to be' soliloquy, for instance.)

More recently Jacques Derrida has adopted and developed the term, and Alan Bass, the English translator of *Writing and Difference*, glosses Derrida's usage of it as follows: 'once a system has been "shaken" by following its totalizing logic to its final consequence, one finds an excess which cannot be construed within the rules of logic, for the excess can only be conceived as *neither* this *nor* that, or both at the same time – a departure from all rules of logic' (Derrida 1978, xvi). For Derrida, according to Bass, this excess is often posed as an aporia.

In the wake of Derrida the term has become more popular as a way of referring to those irresolvable doubts and hesitations which are thrown up by the READING of a TEXT. The term is not normally used (by those who normally use it) in a pejorative sense or to indicate disapproval, but rather to point to SITES within a reader's experience of a text in which he or she is given the freedom to play with the text by the irresolvability revealed at its stress points or fault lines.

Apparatus From the German dramatist Bertolt Brecht's term *Apparat*, meaning an ensemble of CONVENTIONS, conditions, modes of production which lies partly hidden to the GAZE of the participant. Brecht names opera, the stage and the press as examples of an apparatus in his notes to the opera *Aufstieg und Fall der Stadt Mahagonny*. He suggests that such apparatuses 'impose their views as it were incognito' (Willett 1964, 34). The concept is related to Brecht's belief in the need for the artist not just to create new WORKS, but to create new possibilities for the creation, dissemination, and reception of art. Thus he attacks 'a general habit of judging works of art by their suitability for the apparatus without ever judging the apparatus by its suitability for the work' (Willett 1964, 34). John Frow has suggested that the literary system can be seen as such an apparatus, and that, accordingly, social functions need to be understood as a component of textuality (1986, 84).

For the Foucauldian use of this term see *DISPOSITIF*.

Appropriateness conditions See SPEECH ACT THEORY

Appropriation See INCORPORATION

Arbitrary In Linguistics a SIGN is arbitrary if the relationship between it and whatever it stands for or represents is fixed by CONVENTION rather than by any intrinsic or inherent resemblance beyond the scope of the particular sign system to which it belongs. Thus human word language typically rests primarily upon arbitrary signs. The French word *chien*, the German *Hund*, and the English *dog* are all used in their respective languages to represent the same animal; none of these words in spoken or written form resembles a dog independently of the conventions governing the use of the three languages in question. In contrast, speaking onomatopoeic words such as *pop* and *hiss* may be said to produce a noise not dissimilar to the sounds the words are used to represent. As signs, therefore, it can be said that these words are not totally arbitrary. Different writing systems can be more or less arbitrary: *pictographs*, for instance, are less arbitrary than words written in *phonetic script*.

The linked terms *motivated* and *unmotivated*, or *natural* and *conventional* are sometimes used to convey a similar point: a motivated or natural sign is one which is linked to that which it represents by a resemblance or connection existing independent of the conventions of the sign system to which the sign belongs. These terms also have application outside of formal sign systems. A motivated or natural symbol, for example, is a symbol which has some natural, extra-systemic resemblance to or connection with that for which it stands.

Ferdinand de Saussure's insistence upon the arbitrary relationship between SIGNIFIER and SIGNIFIED has been very influential in the present century, partly – it is arguable – as a result of misunderstandings of Saussure's point. Certainly some followers of Saussure have used the arbitrary nature of this relationship (between, on the one hand, a sound image or its written equivalent and, on the other, the concept to which it refers) as a basis for seeing language as a self-enclosed system with no necessary connection with extra-linguistic reality. But this is to misunderstand Saussure's point. As Thomas G. Pavel has observed, the principle of arbitrariness maintains only that there is no motivated link between the conceptual and the phonetic sides of a linguistic sign, 'it does not deny the stability of linguistic meaning, once the semiotic system has been established' (1986, 8).

For the RUSSIAN FORMALIST use of the term *motivation* see FUNCTION.

Archaeology of knowledge The title of an influential book by Michel Foucault in which the author attempts to describe DISCOURSES (as he defines this term), their internal rules and structures, interrelationships, continuities and discontinuities, rules of transformation, and the conditions of their emergence, development and decline. He stresses that he does not want to suggest, by his use of the word archaeology, that he is concerned with something frozen and out of time. He sums up: 'The domain of things said is what is called the *archive*; the role of archaeology is to analyse that archive' (back cover note to Foucault 1972).

For Foucault the *archive* is a particular level that comes between the LANGUE that defines the system of constructing possible sentences, and the *corpus* that passively collects the words that are spoken. The archive is 'a practice that causes a multiplicity of statements to emerge as regular events', it is '*the general system of the formation and transformation of statements*' (1972, 130). The job of uncovering the archive 'forms the general horizon to which the description of discursive formations, the analysis of positivities, the mapping of the enunciative field belongs'. And the term archaeology is given to the totality of all these searches (1972, 131).

Arche-writing Also *archi-trace*, *archi-writing*, *proto-writing*. A term invented by the French philosopher Jacques Derrida and derived in particular from Sigmund Freud's essay 'Note on the Mystic Writing Pad'. The writing-pad in question was a toy sold for children on which messages could be written with a hard stylus, but apparently removed by detaching its double covering sheet from the wax slab on which this rested. What interested Freud was that although this operation rendered the writing in question invisible, it did not remove it utterly. The written message was still there, imprinted on the wax, hidden but not completely erased. Thus the wax base could be compared with the UNCONSCIOUS, from which (as Freud repeated on several occasions) nothing was ever completely erased, while the outer layer of celluloid and translucent waxed paper would accordingly be taken to represent the conscious mind which sends information on to the Unconscious without retaining it.

Moreover, the writing that becomes visible on the pad as a result of the use of the stylus was already there, in the sense that the use of the stylus only makes visible part of the wax block that pre-existed the act of writing. This development of Freud's analogy thus involves a conceptualizing of the unconscious mind as constituted by *writing* (see ÉCRITURE) in the form of an arche-writing or ur-writing in the brain which precedes all physical writing and, even, all speech – both phylogenetically and ontogenetically. As Derrida puts it,

> Writing supplements perception before perception even appears to itself [is conscious of itself]. 'Memory' or writing is the opening of that process of appearance itself. The 'perceived' may be read only in the past, beneath perception and after it. (1978, 224)

From this perspective no perception is virginal or direct, but is given MEANING by a pre-existing arche-writing. The theory has certain points of contact with Noam Chomsky's theory of a Language Acquisition Device (LAD) which, because it must predate the actual learning of any language, has the effect of situating language (again in the form of an ur-language or set of language universals) in the biological brain rather than on a social or CULTURAL terrain.

Derrida takes his use of the word TRACE from the same source in Freud. When the words written on the writing pad are removed, a slight scratch or trace remains on the surface. Freud sees this to be representative of the manner in which a trace 'is left in our psychical apparatus of the perceptions which impinge upon it' (quoted by Derrida 1978, 216). But the perceptions themselves are more than this trace: they are constituted by the relation between this trace and that which makes them visible (in the writing pad, the pressure of the wax slab behind; in the human mind, the coming together of trace and the Unconscious – the two linked by Freud in a passage quoted by Derrida [1978, 225]).

Derrida then moves to apply this to writing (or, perhaps more properly, écriture). Freud 'performs for us the scene of writing', according to Derrida, but Freud's concept of the trace must nevertheless be 'radicalized and extracted from the metaphysics of presence' (1978, 229). For Derrida, the trace would thus be

> the erasure of selfhood, of one's own presence, and is constituted by the threat or anguish of its irremediable disappearance, of the disappearance of its disappearance. An unerasable trace is not a trace, it is a full presence . . .
> (1978, 229)

Architext Following Gérard Genette's book *L'introduction à l'architexte* (1979), that ideal TEXT implied by the generic tradition to which a particular text belongs.

Archive See ARCHAEOLOGY OF KNOWLEDGE

Askesis See REVISIONISM

Aspect According to Mieke Bal the aspects of a given STORY are those traits which are specific to it and which distinguish it from other stories (1985, 7). Gerald Prince, in contrast, defines aspect as the *vision*, or point of view in terms of which a story is presented (1988, 7), while Gérard Genette's definition of the same term is 'the way in which the story is perceived by the narrator' (1980, 29). As with many recent terms within NARRATIVE theory, *aspect* is borrowed from Linguistics, and its use rests upon an argued HOMOLOGY between the STRUCTURE of narratives and the structure of sentences.

What grammarians categorize as aspect is, in Genette's terminology, known as FREQUENCY (1980, 113).

Assimilation In Mikhail Bakhtin's theories of DIALOGUE: the process whereby an individual temporarily adopts the viewpoint or IDEOLOGY of another person, or *assimilates* these to his or her own consciousness. Such assimilation will be more or less complete, more or less whole-hearted, to the extent that the individual's viewpoint or ideology are or are not at odds with those of the interlocutant. The concept can be seen as comparable in certain ways to Samuel Coleridge's 'willing suspension of disbelief', although Bakhtin seems to move more in the direction of 'adoption of belief'.

Aura The German MARXIST Walter Benjamin used the term aura (adj. auratic) to describe the mystical sense that surrounds artistic or ritual objects like a halo, an aura that, according to him, is ultimately destroyed by techniques of mechanical reproduction such as photography. In his essay 'The Work of Art in the Age of Mechanical Reproduction' (included in *Illuminations*, 1973), he writes that

> that which withers in the age of mechanical reproduction is the aura of the work of art. This is a symptomatic process whose significance points beyond the realm of art. One might generalize by saying: the technique of reproduction detaches the reproduced object from the domain of tradition. (1973, 221)

In another essay included in *Illuminations*, 'On Some Motifs in Baudelaire', he further designates aura as 'the associations which, at home in the *mémoire involontaire*, tend to cluster around the object of a perception' (1973, 188). The *mémoire involontaire*, for Benjamin, is taken from the novelist Marcel Proust's distinction between involuntary and voluntary recollection, and rests on the belief (attributed by Benjamin to Freud) that memory fragments are often most powerful and most enduring when the incident on which they are based never entered consciousness (1973, 162). For Benjamin, mechanical reproduction interrupts the process whereby such auratic components enter the involuntary memory and enrich the art-work.

Author Subsequent to the publication of two essays in particular – Roland Barthes's 'The Death of the Author' (1977) and Michel Foucault's 'What is an author?' (1980b) – the term *author* has, from being one of the least problematic of terms, become the SITE of much complex discussion. Clearly on a simple level an author is a person who writes a WORK: Emily Brontë is the author of *Wuthering Heights*. But as Foucault reminds us, one cannot be an author by writing just anything: a private letter may be signed by the person who wrote it, but it does not have an author, nor do we normally speak of the author of a scientific theory.

In other words, the term *author* does more than attach a piece of writing to its individual human origin: it has to be a special sort of writing, and the relation thus posited is more than a certificate of origin. As Foucault puts it, 'The author-function is therefore characteristic of the mode of existence, circulation, and functioning of certain discourses within a society' (1980b, 148). To talk of an author is to appeal to a shared knowledge of these DISCOURSES and of the CONVENTIONS governing their transmission and circulation.

Foucault states that the concept of authorship comes in the train of what he calls 'penal appropriation': it is when writers become subject to punishment for what they have written that works start to acquire authors, and he dates the emergence of authorship at the point at which writers entered the 'system of property that characterizes our society' (1980b, 149). Barthes makes a very similar point, claiming that the author is a modern figure,

> a product of our society insofar as, emerging from the Middle Ages with English empiricism, French rationalism and the personal faith of the Reformation, it discovered the prestige of the individual, of, as it is more nobly put, the 'human person'. (1977, 142–3)

Perhaps the most challenging of Foucault's arguments is that *author* and *living person who wrote the work* are not to be equated, as the author function can give rise to several selves, several subjects, 'positions that can be occupied by different classes of individuals' (1980b, 153). When we talk of an author, in other words, we have in mind a range of characteristic actions and relationships which we do not attribute to every writing individual. What these are is, of course, a complex matter – especially as (according to both Barthes and Foucault), the author function is not a historically stable one.

Barthes in particular is keen (as the title of his essay suggests) to *challenge* the power of the author, a power to which he attributes a range of specifically IDEOLOGICAL functions. For him, to seek to explain a work by reference to the person who wrote it is (by implication) to be in thrall to a pernicious sort of individualism, and to imprison the work in the imagined self of its individual producer. The alternative view involves moving from work to TEXT, seen not as personal statement from author-God but as 'a tissue of quotations drawn from the innumerable centres of culture' (1977, 146). Clearly here the author is seen in representative POST-STRUCTURALIST manner as site rather than originating PRESENCE.

Barthes's view here is intimately tied up with theories of ÉCRITURE, translated rather less than satisfactorily in the English version of his essay as 'writing'. For him, writing

> is the destruction of every voice, of every point of origin. Writing is that neutral, composite, oblique space where our subject slips away, the negative where all identity is lost, starting with the very identity of the body writing. (1977, 142)

A similar theoretical basis underlies Barthes's claim that, for Mallarmé, it is language which speaks, not the author. 'The death of the author' is, then, an aspect of the POSTMODERNIST and POST-STRUCTURALIST attack on *origins*, on the belief that we can explain (or even help to understand) anything by referring it to where we think that it comes from or to any process of cause and effect. Barthes closes his essay by claiming that the birth of the READER, must be at the expense of the death of the author.

A textual critic such as Jerome J. McGann (1983) has drawn attention to the gap between the author seen as sole creator of the work, and the real process of literary composition involving negotiation between historically located individual author and a range of other individuals and institutions – publishers, editors, censors, collaborating friends, critics, and so on.

The terms *implied author* and *implied reader* (see the entry for READERS AND READING) come from Wayne C. Booth's *Rhetoric of Fiction* (1961). An implied author is that sense of the creating author behind a literary work that the reader builds up on the basis of the individual work. The related term *career author* is used by the narratologist Seymour Chatman to denote that sense of a personality or human presence which readers construct from the historical author's (= the author as 'real person') work*s* (not work).

Authorial See NARRATIVE SITUATION

Authoritative discourse See DISCOURSE

Autodiegetic See DIEGESIS AND MIMESIS

Automization See DEFAMILIARIZATION

Autopoiesis Taken by Jerome J. McGann from *Autopoiesis and Cognition: The Realization of the Living*, by Humberto Maturana and Francisco Varela (1980).

McGann suggests that the term can be applied to literary texts, as these are paradigms of the interactive and feedback mechanisms studied by Maturana and Varela. Such mechanisms are distinguished from 'vehicular textual models' whose textual paradigm is one that does not interfere with or 'distort' a message. With vehicular textual models one can easily distinguish vehicle of transmission from message, whereas this is not the case with *autopoiesis* (McGann 1991, 11).

Compare the discussion in the entry for CODE.

Avant-garde See MODERNISM AND POSTMODERNISM

B

Backgrounding See DEFAMILIARIZATION

Base and superstructure (also basis and superstructure) Central to the traditional MARXIST analysis of society and history is the analytical distinction between base and superstructure. The most famous statement of this distinction is to be found in Karl Marx's *A Contribution to the Critique of Political Economy*, in which he states that

> In the social production of their existence, men inevitably enter into definite relations, which are independent of their will, namely relations of production appropriate to a given stage in the development of their material forces of production. The totality of these relations of production constitutes the economic structure of society, the real foundation, on which arises a legal and political superstructure and to which correspond definite forms of social consciousness. The mode of production of material life conditions the general process of social, political and intellectual life. It is not the consciousness of men that determines their existence, but their social existence that determines their consciousness. (1971, 20-21)

On the basis of this position traditional Marxists have distinguished between elements in society which, with regard to their emergence and their historical effect, are either primary or secondary. On the one hand the *economic structure of society, the real foundation* (seen as primary), and on the other the *superstructure* (seen as secondary). Along with law and politics (mentioned by Marx), the superstructure has generally been taken to include other cultural and intellectual phenomena such as (in some accounts) literature.

Such an analytical position leads inexorably to the view that to understand literature one must understand the primary phenomenon of which it is the secondary reflection: the economic base of society. Thus Terry Eagleton, has suggested that 'Marxist criticism is part of a larger body of theoretical analysis which aims to understand *ideologies*', and he locates IDEOLOGIES firmly within the superstructure (1976b, viii, 4).

We must however remember that the 'material forces of production' do not just consist of machines and factories, but also of human skills. To take a relevant example: a wholly or largely illiterate society is less able to produce wealth in the modern world than one possessed of a literate workforce. Clearly, then, literacy is an element in the forces of production. Following Marx's comment in the first volume of *Capital* that the difference between the worst architect and the best of bees is that 'the architect raises his structure in imagination before he erects it in reality' (1970a, 178), it can be inferred that human imagination is implicated in the productive forces. Thus if literature contributes to the development of literacy and imagination it is hard to relegate it lock, stock and barrel to Marx's superstructure.

More recently Richard Harland has coined the term *superstructuralism* to refer a range of recent theorists who 'invert our ordinary base-and-superstructure models until what we used to think of as superstructural actually takes precedence over what we used to think of as basic' (1987, 1–2). Harland has in mind not just STRUCTURALISTS and POST-STRUCTURALISTS but also SEMIOTICIANS, Foucauldians and Althusserian Marxists.

Belatedness See REVISIONISM

Binary / binarism Binary oppositions form the basis of DIGITAL systems of communication, and involve the reduction of *continuous variations* to *discrete either/or distinctions*. This process is fundamental to human language: the classic example involves the colour spectrum which, although continuous in nature, is broken up into discrete segments which are given separate colour terms in various human languages.

Binary distinctions are fundamental to much of modern Linguistics, and Jonathan Culler quotes the linguist Charles Hockett's remark that 'If we find continuous-scale contrasts in the vicinity of what we are sure is language, we exclude them from language' (Culler 1975, 14).

STRUCTURALISM has adopted this principle of modern Linguistics and has given it a central position in structuralist theory. Structuralist analysis typically searches for hierarchical strings of binary oppositions in the material or TEXT under investigation: a classic example is indicated in the title of Claude Lévi-Strauss's *The Raw and the Cooked*. Jacques Derrida's coinage DIFFÉRANCE is a direct descendant of the structuralist emphasis upon binary oppositions.

See the more detailed entry under DIGITAL AND ANALOGUE COMMUNICATION.

Body For many recent theorists the body is no longer merely a purely physical system the study of which can safely be left to the medical profession. Instead, the body is also a concept or set of ideas which are seen to be implicated in, and already in part constructed by, the non-physical: IDEOLOGY and history, for example.

Jacques Lacan sees the infant's changing attitude to his or her body as crucial to progress through the MIRROR STAGE. Michel Foucault has suggested that instead of power writing its version of truth directly on the body through torture, this function has been taken over in the modern world by schools and prisons which attempt to reform (see his *Discipline and Punish: The Birth of the Prison* [1979]).

Moving in a rather different direction, many recent FEMINISTS have stressed the importance for women of reinscribing the body in their writing (see the entry for ÉCRITURE FÉMININE). This is a different direction, because (at least in some accounts of écriture féminine) this treats the body more as source than as site.

But what perhaps all of these accounts do have in common is an attempt to historicize the body, to come to terms with the fact that the body is conceptualized differently in a range of CULTURES and historical periods.

Bottom up (perspective) See SOLUTION FROM ABOVE / BELOW

Bricoleur In his *The Savage Mind* Claude Lévi-Strauss distinguishes between the SIGN systems of modern man and those of primitive man. Modern man, like an engineer, makes use of specialized and custom-made tools and materials, whereas primitive man resembles an odd-job man or *bricoleur*, who makes use of those odds and ends of material which he has to hand to construct pieces of *bricolage*.

> The characteristic feature of mythical thought is that it expresses itself by means of a heterogeneous repertoire which, even if extensive, is nevertheless limited. It has to use this repertoire, however, whatever the task in hand because it has nothing else at its disposal. Mythical thought is therefore a kind of intellectual 'bricolage' . . . (1972, 17)

As a literary example of a *bricoleur* Lévi-Strauss points to the CHARACTER Wemmick, from Charles Dickens's *Great Expectations*, because Wemmick creates a sort of MYTH from the raw materials at hand: the parts of his suburban villa are mythically transformed into a castle (1972, 17 and 150n).

Lévi-Strauss argues, accordingly, that the sign SYSTEMS of primitive man tend to be less ARBITRARY or unmotivated, more MOTIVATED than those of modern man. This, he continues, reflects back upon systems and habits of thought, as whereas modern man is able to separate the abstract sign systems he uses from the CONCRETE realities to which they refer, for primitive man the motivated nature of his sign systems means that reality and sign systems are perceived as more interdependent and involved in each other. For Lévi-Strauss this leads, paradoxically, to the position that the thought of primitive man is both concrete and abstract, and that separating the concrete from the abstract is much easier for modern man.

Jacques Derrida has attempted to DECONSTRUCT the opposition between *bricoleur* and engineer by claiming that if by *bricolage* one refers to the necessity of borrowing one's concepts from 'the text of a heritage which is more or less coherent or ruined, [then] it must be said that every discourse is *bricoleur*' (1978, 285). This clearly attracts Derrida because it is easier to argue against the unity or absolute source of a *bricolage* than of some other TEXTS – unless, of course, *all* texts are *bricolages* in which case no text can be possessed of such a unity or absolute source – or PRESENCE, CENTRE, or any other hierarchically organizing discipline.

Those writing about the phenomena of MODERNISM and POSTMODERNISM have also considered the notion of *bricolage* to be useful, for it seems to have a relevance to the typically (post)modernist incorporation of 'found' objects together in new STRUCTURES which celebrate their diversity and heterogeneity.

Brisure See HINGE

C

Cancelled character A term coined by Brian McHale to describe a technique whereby a literary character is exposed as textual function and no longer seen as 'integral creature' possessed of self-identity. McHale gives as example Tyrone Slothrop, from Thomas Pynchon's *Gravity's Rainbow*, who

> demonstrates this textualized concept of character: beginning as at best a marginal self, he literally becomes *literal* – a congeries of *letters*, mere words. The zone in which he is lost and scattered is not only a heterotopian projected space but, literally, a space of writing, and his disassembly 'lays bare' the absorption of character by text. (1987, 105)

According to McHale, this example lays bare what is more or less implicit in a range of other POSTMODERNIST characters: 'the ineluctable writtenness of character' (1987, 105). The technique is arguably implicit in MODERNIST portrayals of character, such as that of Klamm in Kafka's *The Castle*.

See also CHARACTER; CHARACTER ZONE.

Canon The term originates in debate within the Christian Church about the authenticity of the Hebrew Bible and books of the New Testament. That which was termed canonical was accepted as having divine authority within the Church, while writing of no, or doubtful, authority was termed apocryphal.

Thus the Protestant canon and apocrypha differ slightly from those of the Catholic Church.

In ecclesiastical use, then, *canon* refers both to *origin* and *value*. By extension the term in literary-critical usage came to be applied (i) to WORKS which could indisputably be ascribed to a particular AUTHOR, and (ii) to a list of works set apart from other literature by virtue of their literary quality and importance. By the middle of the present century such a decision was to a large extent decided institutionally: just as the church (or different churches) decided upon the Biblical canon, so the universities decided of which literary works the literary canon consisted. This is not to say that there were no disagreements concerning the canon: many of the disputes in which F. R. Leavis was involved during the 1930s and 1940s were essentially concerned with the canon. Leavis's view of the canon was highly restrictive: his novelistic Great Tradition consisted essentially of some of the work of just four novelists.

When certain literary critics began to speak of canon*s* (cf. Robert von Hallberg 1985) the move had important theoretical implications. After all, although the Protestant and Catholic Churches may disagree about canon and apocrypha, they all agree that only one of them can be correct. But were a church suddenly to state that no one canon had absolute authority, as canons only represented the needs and viewpoints of particular churches, then the idea of the canon as somehow linked to divine origin and authority would inevitably be opened to question.

This at any rate is what appears to have happened in literary-critical circles. When FEMINIST critics started to construct a rival canon or canons, not always as a *replacement* for the 'official' canon but also as an *alternative* to it, then this struck at the *claim to universality* behind the idea of a single canon. For, in the traditional sense, if there were several canons then there was no canon.

For Mikhail Bakhtin, *canonization* is a process towards which all literary genres have a tendency, in which temporary norms and CONVENTIONS become hardened into universal ones so that evaluations too are considered to reflect universal rather than culture- or time-bound values. Thus Bakhtin suggests that in its own time the HETEROGLOSSIA in a novel will be readily recognized, but as the work recedes in time this heteroglossia is more and more obscured by the process of canonization, which standardizes and reduces the ways in which the work can be read (1981, 417–18).

John Guillory has argued, interestingly, that the establishment and modification of the canon are intimately connected with the pedagogy of language teaching, and that those works that have been canonized have been chosen – at least in part – because they have been found suitable for the more or less sophisticated needs of language teaching at different educational levels and at different periods of history (Lentricchia & McLaughlin 1990, 240–43).

A few critics have attempted – covertly or overtly – to restrict the term *literature* to works within the canon, thus necessitating the use of other terms to describe non-canonical poetry, fiction and drama. The term *paraliterature* has been coined to describe such work which is seen to be 'literary' in a broad sense but non-canonical, such as crime fiction, romantic fiction, the sort of poetry published in mass-circulation magazines, and so on. The emergence of such new terms is associated with a historical development which alters the scope of 'literature' from that of being a broad descriptive term to a far narrower, honorific term favoured by critics such as F. R. Leavis. In recent years, however, a counter-movement to expand the scope of this term and to reject what have become to be seen as élitist attempts to narrow its scope has become apparent.

Career author See AUTHOR

Carnival In the writings of Mikhail Bakhtin the significance of the carnival in the Renaissance and Middle Ages assumes a representative importance, indicative of a particular form of POPULAR counter-culture. During this period, according to Bakhtin, 'A boundless world of humorous forms and manifestations opposed the official and serious tone of medieval ecclesiastical and feudal culture' (1968, 4). No matter how variegated and diverse these forms and manifestations they nonetheless, claimed Bakhtin, belonged to a single CULTURE of folk carnival humour (1968, 4). Bakhtin sees this culture as existing on the borderline between art and life, 'life itself, but shaped according to a certain pattern of play' (1968, 5). There is no firm distinction between actors and spectators, and during the period of the carnival it embraces all the people and there is no life outside of it.

By extrapolation Bakhtin dubs as *carnivalesque* all manifestations of a comparable counter-culture which is popular and democratic, and in opposition to a formal and hierarchical official culture. Important here is the idea of unity-in-diversity, of the heterogenic unison or POLYPHONY of the many voices that make up the carnival. In more recent usage, then, *carnival* and *carnivalesque* refer to traditional, often spontaneous, cultural phenomena which, although assuming many different surface forms, are nonetheless deemed to express a common oppositional or counter-culture. For Bakhtin, a sure distinguishing feature of the carnivalesque is laughter, which is never allowed in official celebration – something that Bakhtin, writing in the Soviet Union under Stalin, knew all about.

In his study of Dostoevsky Bakhtin suggests that the influence of carnival is responsible for a set of 'genres of the serio-comical', and that these have three main defining features. First, that 'their starting point for understanding, evaluating, and shaping reality, is the living *present*'; second that they do not rely on *legend* but, *consciously*, on *experience* and *free invention*; and third, that they are deliberately 'multi-styled and hetero-voiced' (1984, 108).

To what extent such manifestations offer genuine opposition to the official culture has been the topic of some debate. Is the carnival like the licensed fool, the permitted safety-valve which reinforces the culture it mocks and parodies? Or is it the unassailable SITE to which a genuine oppositional force retreats when more overt challenges to official culture (and the political power that lies behind them) are not possible? (See the entry for INCORPORATION.)

Bakhtin argues that the carnival is the place for working out,

> in a concretely sensuous, half-real and half-play-acted form, a *new mode of interrelationship between individuals*, counterposed to the all-powerful socio-hierarchical relationships of non-carnival life. (1984, 123)

He further suggests that the *eccentricity* legitimized by carnival permits 'the latent sides of human nature to reveal and express themselves' (1984, 123), a suggestion that obviously has a bearing upon his interest in Dostoevsky's use of eccentrics in his fiction.

Such debates have a clear relevance to current discussions of popular culture and popular literature, and to debates about the distinction between 'of the people' and 'for the people', or folk and commercial culture.

Catalyses See EVENT

Centre In the work of Jacques Derrida the term *centre* is used to represent 'a point of presence, a fixed origin' (1978, 278) which imposes a limit on the play of the STRUCTURE in which it is found or placed. Following Derrida, then, much of the energy of DECONSTRUCTIVE criticism is directed towards freeing structures from the tyranny of whatever centre or centres to which they are seen to be subject. Derrida also uses a range of other terms, including *origin*, *end*, *arche*, and *telos* as roughly equivalent to centre.

Building on such usages, Vincent Crapanzano explains that his own use of the term posits

> an image, an event, or even a theoretical construct functioning as a nucleus or point of concentration that holds together a particular verbal sequence. The center gives coherence, a semblance of order at least, to what would otherwise appear to be a random, meaningless sequence of expressions. (1992, 28)

Crapanzano argues that centring can operate both precursively and recursively – that is, that a centre can condition or determine the meaning of that which precedes and also that which follows the centre.

When critics talk of decentring the SUBJECT they should be understood to be extending this argument to the human subject. The human subject is thereby denied a unity that is underwritten and orchestrated by a controlling centre which, like an all-powerful micro-chip in a super-computer, brings the whole system into synchrony with and through its all-pervasive PRESENCE and discipline. In the light of such an approach the human subject becomes SITE rather than point of origin, and a site, moreover, on which unrelated campers come and go (and sometimes fight) rather than one united by an all-powerful scout-master.

A related term is 'the god trick', which refers to the way in which God (or any divinity) can be used as a centre to fix the meaning of all in the experience of the believer. Normally the divinity referred to is a metaphoric rather than a literal one: the term tends to be used in a sarcastic or dismissive manner.

See also COPERNICAN REVOLUTION; LOGOCENTRIC; TRANSCENDENTAL SIGNIFIED.

Centrifugal / centripetal Mikhail Bakhtin uses these terms which describe the impulse either outwards or inwards, from or towards the centre, to refer to social and IDEOLOGICAL rather than physical forces. (Bakhtin's use of the word CENTRE is different from that of Jacques Derrida: see the separate entry for this term.)

Both of these forces, for Bakhtin, are to be found in language: there is the impulse and pressure towards a standardization imposed and maintained by a central authority, and at the same time the urge towards diversity and POLYPHONY. Every language, he claims in his essay 'Discourse in the Novel',

> participates in the 'unitary language' (in its centripetal forces and tendencies) and at the same time partakes of social and historical heteroglossia (the centrifugal, stratifying forces). (1981, 272)

For him certain literary genres have a centripetal force, driving READERS towards a centre of conformity and uniformity, whereas others urge people away from conformity and towards diversity and heterogeneity. He accords poetry an essentially centripetal tendency, while the novel is granted the opposite, a centrifugal force. Not surprisingly, he argues that the novel flourishes during times of diversity and the slackening of central control.

Character The concept of literary character has had a generally hard time in Anglo-American and European critical theory of the present century. The New Critics and F. R. Leavis were united in their disapproval of A. C. Bradley's alleged treatment of Shakespearian (and other) characters as if they were real people, and Alan Sinfield has reminded us that G. Wilson Knight set aside 'the character' as category of analysis because of his belief that each play was a visionary whole, 'close-knit in personification, atmospheric suggestion, and direct poetic-symbolism' (Sinfield 1992, 56–7, quoting from Knight's *The Wheel of Fire*).

Subsequent critical approaches have generally shown little more enthusiasm for the concept or category. This has much to do with the anti-HUMANISM of much recent theory: humanism has been accused of privileging a human essence and of situating this within the bourgeois SUBJECT. The attack on literary character can thus be seen as an aspect of the anti-individualistic bias of much recent theory. To quote a representative example: in his *The Postmodern Condition: A Report on Knowledge* Jean-François Lyotard comments that

> A *self* does not amount to much, but no self is an island; each exists in a fabric of relations that is now more complex and mobile than ever before. Young or old, man or woman, rich or poor, a person is always located at 'nodal points' of specific communication circuits, however tiny these may be. Or better: one is always located at a post through which various kinds of message pass. (1984, 15)

Even within the study of NARRATIVE, no doubt as a result of the STRUCTURALIST foundations of NARRATOLOGY, the individual character – whether traditional or modernist – is rigorously DECONSTRUCTED. Gerald Prince's definitions of the term 'character' – 'An existent endowed with anthropomorphic traits and engaged in anthropomorphic actions; an actor with anthropomorphic attitudes' (1988, 12) – almost leads one to feel that considering a literary character as in some way equivalent to a real human being is akin to talking to one's cat. But just as people have gone on talking to their cats so too critics and READERS have gone on treating literary characters as in certain ways equivalent to human individuals.

Thus Alan Sinfield's essay 'When is a Character Not a Character? Desdemona, Olivia, Lady Macbeth and Subjectivity', which is to be found in his book *Faultlines*, clearly finds the category less than fully played out. Indeed, by arguing that Desdemona 'is a disjointed sequence of positions that women are conventionally supposed to occupy', and that she 'has no character of her own; she is a convenience in the story of Othello, Iago, and Venice' (1992, 53, 54), he forces the reader to compare her with those others in the play who, he suggests, *are* so possessed of a character.

See also CANCELLED CHARACTER; CHARACTER ZONE.

Character zone According to Mikhail Bakhtin, character zones

> are formed from the characters' semi-discourses, from various forms of hidden transmission for the discourse of the other, by the words and expressions scattered in this discourse, and from the irruption of alien expressive elements into authorial discourse (ellipsis, questions, exclamation). Such a zone is the range of action of the character's voice, intermingling in one way or another with the author's voice. (Quoted in Todorov 1984, 73)

In other words, the READER of a novel builds up a picture of the CHARACTER'S identity not just from direct descriptions of his or her actions or from 'transcriptions' of his or her speech, but from a wider zone of verbal implication.

See also CANCELLED CHARACTER; CHARACTER ZONE; ERASURE (erased character).

Chinese box narrative See FRAME

Chronological deviation See ANACHRONY

Chronotope A coinage of Mikhail Bakhtin's to designate 'the distinctive features of time and space within each literary genre' (Todorov 1984, 83). According to Bakhtin he took the term from mathematical biology, but introduced it into literary studies in a metaphorical sense (Bakhtin 1981, 234). In his essay on the *Bildungsroman*, for example, Bakhtin suggests that the 'local cults' associated with particular literary works and linked to specific geographical locations in the mid eighteenth century 'attest above all to a completely *new sense of space and time* in the artistic work' (1986, 47).

Circulation A term favoured by the NEW HISTORICIST writer Stephen J. Greenblatt to describe the manner in which CULTURAL artifacts or IDEOLOGICAL meanings (for example) are transmitted from place to place by means of 'practical strategies of negotiation and exchange' (1990, 154). Greenblatt acknowledges the influence of Jacques Derrida in his use of the term, and what seems to be important is the way in which it presupposes a sequence of *transformations* of the artifact or MEANING as it is passed on by means of the aforesaid strategies. (Thus an object in a museum has been transformed from what it was when it was, for example, an item in daily use.)

See also RESONANCE, along with the other cross-references given in the entry for NEW HISTORICISM AND CULTURAL MATERIALISM.

Cleavage See HINGE

Clinamen See REVISIONISM

Closed texts See OPEN AND CLOSED TEXTS

Closure All literary WORKS come to some sort of close or conclusion, but such conclusions do not always provide the same sense of satisfaction or inevitability. According to Barbara Herrnstein Smith, closure 'may be regarded as a modification of structure that makes *stasis*, or the absence of further continuation, the most probable event'. She further adds that closure allows the READER to be satisfied by the failure of continuation or, to put it another way, it creates 'the expectation of nothing' in the reader (1968, 34).

A lack of closure is frequently associated with MODERNIST or experimental art, including literature. Modernist writing often seems to challenge those literary CONVENTIONS which arouse certain expectations in the reader regarding what is acceptable as an ending for a literary work. Such works are often described as *open-ended* or lacking closure. This can refer not just to matters of STORY or PLOT, but also to aesthetic and IDEOLOGICAL issues.

Consonant and *dissonant* closure occur when the ending of a literary work either confirms and underwrites, or challenges and destabilizes, what has gone before.

Code The influence of Linguistics and SEMIOTICS has led to an increased recourse to the term *code* on the part of literary critics and theorists during the past two decades. This may be at least in part because the term implies that writer and READER are linked by their common possession of a set of CONVENTIONS governing systematic transformations, an implication which appeals to many contemporary theorists interested in issues raised by the sociology of literature and by the concept of LITERARINESS. The term also, of course, suggests that the literary work contains that which is hidden to those not possessed of the right code-book. (In this context, see also INTERPRET[AT]IVE COMMUNITY.)

In one of the more influential examples of the development of a theory of literary codes, Roland Barthes has suggested five codes of reading which allow readers to recognize and identify elements in the literary WORK and to relate them to specific FUNCTIONS. The five codes are as follows.

The *proairetic code* controls the manner in which the reader constructs the PLOT of a literary work.

The *hermeneutic code* involves problems of interpretation, particularly those questions and answers that are raised at the level of plot.

The *semic code* is related to the TEXTUAL elements which develop the reader's perception of literary CHARACTERS.

The *symbolic code* governs the reader's construction of symbolic MEANINGS.

The *referential code* is made up by textual references to CULTURAL phenomena. (Based on Barthes 1990, 19–20; the codes are also discussed in Barthes 1975).

Few critics apart from Barthes himself have put these terms to extended literary critical use, although one who has is Robert Scholes in his book *Semiotics and Interpretation.* The book contains a chapter dealing with James Joyce's short story 'Eveline', in which Barthes's various codes of reading are pressed into explanatory service (1982, 99–104).

Elsewhere in his work Barthes uses references to a variety of other codes. Thus in the course of two pages of his 'Textual Analysis of Poe's "Valdemar"' (1981a) there are references to the metalinguistic code, the socio-ethnic code, the symbolic code, the social code, the narrative code, the cultural code, the scientific code, the scientific-deontological code and the symbolic code. The use of the word code by literary critics has an important relevance with regard to debates about interpretation. To decode a SIGN or sequence of signs is to return to a MEANING or MESSAGE that pre-existed sign or signs, whereas to engage in interpretation (at least according to certain theorists) is to generate something new.

Given that a key dispute between rival theories of interpretation centres on whether or not the interpreter is in part creator of something original, use of terms such as *code* or *decoding* tends to ally one with those who do not see interpretation as a creative activity. Jonathan Culler thus raised an important and early objection to the use of these terms when he remarked that 'listeners interpret sentences rather than decode them' (1975, 19). Although his objection concerns the study of natural languages, it would appear to have even greater force in the field of literary criticism.

Umberto Eco has coined the term *overcoding* to refer to a process of meta-communication about an expression. According to him, overcoded rules tell the reader whether a given expression is used rhetorically: thus the opening statement 'Once upon a time' is deemed by him to convey the facts (i) that the reported events take place in an indefinite non-historical epoch, (ii) that they are not 'real', and (iii) that the speaker wants to tell a fictional story (1981, 19).

The term *delayed decoding* was coined by Ian Watt in his *Conrad in the Nineteenth Century* (1980) to describe a particular impressionist technique of Joseph Conrad's whereby the experiences of a character who understands what is happening to him or her only in the course of, or after, these experiences, is recreated in the reader. Thus in *Heart of Darkness* we share Marlow's belief that lots of little sticks are dropping on the ship, up to the point when Marlow realizes that the 'sticks' are in fact arrows, and that the ship is being attacked.

See also HERMENEUTICS; FUNCTIONS OF LANGUAGE.

Coloured narrative See FREE INDIRECT DISCOURSE

Commissive See SPEECH ACT THEORY

Commutation test Commutation involves the substitution of one thing for another; the commutation test involves investigating the function or significance of one thing by substituting it with another.

In literary analysis one might, accordingly, substitute a first-person NARRATOR for an omniscient one, or substitute Direct Speech for FREE INDIRECT DISCOURSE, in order to assess what the substituted element contributed to the passage or WORK in question.

Competence and performance A distinction introduced into Linguistics by Noam Chomsky which separates those language rules internalized by a native speaker which enable him or her to generate and understand grammatically correct sentences (competence), from the actual generation of particular correct sentences by such a speaker (performance). Competence also enables a native speaker to recognize whether or not a particular sentence is or is not grammatically well-formed.

The distinction has been extended to literary criticism by a number of theorists who have sought to draw an analogy between the internalized rules of a language and the internalized rules or CONVENTIONS which enable competent READERS to read and understand literary WORKS. The analogy has the virtue of reminding us that there is a difference between literacy and the ability satisfactorily to read and respond to literary works, but it has some shortcomings too. First, that whereas Chomsky's competence is mainly concerned with the *generation* of correct sentences, the posited literary competence has often been concerned mainly or exclusively with the *reading or reception* of literary works. Second, that literary competence seems to be CULTURALLY rather than biologically planted in the human individual (there is no literary equivalent to Chomsky's genetically transmitted 'Language Acquisition Device'). And third, literary competence is not universally present in adult human beings. We do not have to go to university to learn how to speak.

Umberto Eco suggests a formulation which may go some way to resolving these problems, when he argues that the well-organized TEXT both presupposes a model of competence coming from outside of itself, but that it also attempts to build up such a competence by merely textual means (1981, 8).

Compare attempts to apply the distinction between LANGUE AND PAROLE to literature (see the separate entry on these terms).

Conative See FUNCTIONS OF LANGUAGE

Concretization *Concrete* was a favourite word of the New Critics and of F. R. Leavis, used honorifically to distinguish literature (mostly poetry) which called particulars to mind, often by means of direct evocations of the recalled testimony of the senses. Take for example Leavis's discussion of four lines from John Keats's 'Ode to Melancholy':

> Then glut thy sorrow on a morning rose,
> Or on the rainbow of the salt sand-wave,
> Or on the wealth of globed peonies;
> Or if thy mistress some rich anger shows . . .

Leavis comments: 'That "glut", which we can hardly find Rossetti or Tennyson using in a poetical place, finds itself taken up in "globed", the sensuous concreteness of which it reinforces; the hand is round the peony, luxuriously cupping it' (1964, 214).

In Roman Ingarden's *Das Literarische Kunstwerk* (published in German in 1931 and in English translation as *The Literary Work of Art* in 1973) we find a use of *concretize* as a verb which has exerted considerable influence subsequent to the appearance of this work in English. According to Ingarden there are two types of concretization if one speaks purely ontically (i.e. 'of or having real being or existence'): on the one hand, 'the purely intentional concretization, ontically heteronomous in form and relative to the subjective operation' and on the other hand, 'the objectively existing concretization, characteristic, in form, of the respective ontic sphere and thus, in a state of affairs that exists in the real world, in the form of an ontically autonomous realization of the corresponding essences or ideas' (1973, 162). The distinction is, it will be seen, a complex one, but it allows Ingarden to talk about the way in which the literary WORK is concretized by being read. Central to his argument is that because literary works typically contain 'spots of indeterminacy', 'the reader usually *goes beyond* what is simply presented by the TEXT (or projected by it) and in various respects *completes* the represented objectivities, so that at least some of the spots of indeterminacy are removed and are frequently replaced by determinacies that not only are not determined by the text but, what is more, are not in agreement with the positively determined objective moments' (1973, 252).

In Brian McHale's account,

> The complexity of the literary artwork, [Ingarden] tells us, lies first of all in its being *heteronomous*, existing both autonomously, in its own right, and at the same time depending upon the constitutive acts of consciousness of a reader. (1987, 30)

The PRAGUE SCHOOL theorist Felix Vodička adopts and develops Ingarden's category in his essay 'The History of the Echo of Literary Works'. For Vodička the study of the concretization of past and present literary works involves 'the study of the work in the particular form in which we find it in the conception of the period (particularly its concretization in criticism)' (1964, 73). Vodička stresses the key point that

> since there is . . . no single correct esthetic norm, there is also no single valuation, and a work may be subject to multiple valuation, during which its shape in the awareness of the perceiver (its concretization) is in constant change. (1964, 79)

Ingarden's 'spots of indeterminacy' contribute to some of Wolfgang Iser's theories about reader creativity; see the entry for PHENOMENOLOGY.

See also ACTUALIZATION; NORM.

Condensation and displacement According to Freud a comparison of the dream-content with the dream-thoughts reveals that 'a work of *condensation* on a large scale has been carried out. Dreams are brief, meagre and laconic in comparison with the range and wealth of the dream-thoughts' (1976, 383). In other words, a large amount of MEANING is *condensed* into a relatively small size by making individual signs or images signify more than one thing: the dream-thoughts are thus OVERDETERMINED. Freud's interpretations accordingly involve a work of unpacking; out of a single scene or figure in a dream a number of different meanings can be salvaged. As Freud also points out, the interpretation of a dream may occupy six or twelve times as much space when written out as does the written account of the dream itself (1976, 383). Dreams involve extreme *concentration* of meaning. Freud suggests a number of ways in which the process of condensation can be carried out – by means of collective figures (one known person in a dream standing for a number of different people, for example), of composite structures (a dream person who combines the appearance and characteristics of a number of different people, for example), or by verbal means (puns, for instance, which unite a number of different words). All of these argued techniques have been of interest to literary critics, and particularly critics of poetry, who have found points of comparison with the concentrated nature of the poetic image or symbol, and indeed many aspects of Freud's dream analyses are remarkably similar to, for example, aspects of New Critical analyses of lyric poems.

For Freud, condensation is typically associated with *displacement*, and he argues that '*Dream-displacement* and *dream-condensation* are the two governing factors to whose activity we may in essence ascribe the form assumed by dreams' (1976, 417). Freud sees displacement as a means whereby censorship is outmanoeuvred; for example, if a person cannot consciously admit his or her hatred of another as a result of the operation of the censor, this hatred may be transferred to something associated with the person in question, *displaced* from one object protected by the censor to another one about which the censor is unconcerned. This again has been of interest to literary critics who have used the concept (often in association with that of condensation) to explore the way symbolism functions in literary WORKS.

Linguisticians use the term DISPLACEMENT in a rather different way: see the separate entry. See also the discussion of *metaphor* and *metonymy* in the entry for SYNTAGMATIC AND PARADIGMATIC: following Lacan, a number of commentators have observed that the condensation-displacement distinction has much in common with Roman Jakobson's distinction between metaphor and metonymy (see the discussion in Scholes 1982, 75–6).

In NARRATOLOGY, *condensation* is sometimes used to indicate that a passage of Indirect or FREE INDIRECT DISCOURSE condenses the (implied) utterances of many different people, perhaps made at many different times, into one represented utterance.

Conjuncture See MOMENT

Connotation and denotation The two terms distinguish between two forms of REFERENCE: the word 'military' as defined in the dictionary involves that which is connected to armies or soldiers (denotation), but it carries with it a range of associations which spring to the minds of those who share a common CULTURE – uniforms, marching, discipline, force, masculinity, rigid collectivity – and these are the *connotations* of the word. Denotations are almost invariably more fixed than connotations, changing only over much longer periods of time than it takes for a word's connotations to alter.

Roland Barthes points out in the introductory comments to his *S/Z* that the precise relationship between connotation and denotation is a matter of dispute. There are thus those for whom denotation is primary and connotation either secondary or non-existent. Some ('the philologists, let us say')

> declaring every text to be univocal, possessing a true, canonical meaning, banish the simultaneous, secondary meanings to the void of critical lucubrations. On the other hand, others (the semiologists, let us say) contest the hierarchy of denotated and connotated; language, they say, the raw material of denotation, with its

> dictionary and its syntax, is a system like any other; there is no reason to make this system the privileged one, to make it the locus and the norm of a primary, original meaning . . . (1990, 7)

A number of recent theorists have suggested that there are similarities between the connotation/denotation distinction and the METONYMY/METAPHOR distinction. This is because both connotation and metonymy involve relations of CONTIGUITY, whereas denotation and metaphor involve relations mediated by CONVENTION. (The difference is, however, that whereas denotation involves UNMOTIVATED relations, metaphor normally relies upon MOTIVATED relations, or upon some aspect of similarity which is external to the system of MEANING involved. Warmongers and peacemakers bear some extra-linguistic resemblances to hawks and doves, whereas the denotation of 'war' as 'armed conflict' is specific to the English language and is thus unmotivated.)

Connotation and denotation are not exclusively linguistic phenomena: the sign of the cross, uniforms, expressive bodily actions, representations of the physical landscape – all these can have both connotations and denotations.

See also SYNTAGMATIC AND PARADIGMATIC.

Consonant closure See CLOSURE

Consonant psycho-narration See FREE INDIRECT DISCOURSE

Constatives See SPEECH ACT THEORY

Contact See FUNCTIONS OF LANGUAGE

Containment See INCORPORATION

Context See FUNCTIONS OF LANGUAGE

Contiguity Literally, the relation of touching or adjoining. METONYMY is defined as that relation based on contiguity, and has led to important theoretical work in both SEMIOTICS and Literary Criticism. The Freudian concept of DISPLACEMENT also relies to some extent on relations of contiguity: thus my fears about my boss may be displaced in a dream on to a fear of the colour blue if that is the colour of the suits he wears. The process of displacement follows a path based on the contiguous relation: boss → blue. See also CONNOTATION AND DENOTATION for discussion of associated concepts.

Mikhail Bakhtin distinguishes between relations of contiguity and those of mutual REFLECTION; merely contiguous lives, he writes,

> are self-enclosed and deaf; they do not hear and do not answer one another. There are not and cannot be any dialogic relationships among them. They neither argue nor agree. (1984, 70)

Convention In traditional usage, either the allowances which are made by READERS and audiences and required by certain genres, or the framework of formal requirements imposed by the same genres or sub-genres. Thus it is a theatrical convention that the actors normally face towards the audience, while other conventions place a certain pressure on both poet and reader with regard to a poem's formal STRUCTURE.

Recent literary and other theory has been much occupied with the larger extent to which systems of rules or conventions underlie the production or recognition of SIGNS and MEANING. A key rôle in the more general concern with this issue has been played by STRUCTURALIST theorists and critics. As Jonathan Culler has put it,

> Structuralism is thus based . . . on the realization that if human actions or productions have a meaning there must be an underlying system of distinctions and conventions which makes this meaning possible. (1975, 4)

Generally speaking, a sign that derives its force from a system of conventions is said to be *unmotivated*, whereas one which has a force independent of any such agreed or accepted conventions is said to be *motivated*.

'Agreed or accepted' is important: conventions may be technically *artificial*, that is to say they may be drawn up, agreed upon, and abided by on the basis of conscious human planning and acceptance. Alternatively, they may be more *natural*, growing in a more unplanned manner, as particular tasks require a set of rules to enable and standardize communication.

See also CONVENTIONALISM, REGISTER.

Conventionalism Terry Lovell suggests that the main thrust of the conventionalist attack on empiricism is that the 'neutral observation language' posited by empiricism cannot be found, and that 'the concepts in terms of which experience is ordered and recorded are not and cannot be theory-neutral' (1980, 14–15). She cites Thomas Kuhn (see the entry for PARADIGM SHIFT) as an exponent of 'flamboyant conventionalism'; according to her he

> declared that all languages of observation and experience are theory-impregnated. He contended that sense perception itself depends upon theory, so that the way in which we perceive the world, the sensations and experiences we have, depend on the theoretical presuppositions we bring to it. (1980, 15)

As she points out, this means that knowledge cannot be validated by experience because the very terms of our experience presuppose certain knowledge. This means that people with different knowledge are cut off from communication with one another, and indeed Kuhn's position has been criticized for positing a set of *paradigms* which are self-enclosed and cut off from one another. As Lovell puts it, 'Because the terms of rival paradigms are incommensurable, reality becomes a function of the paradigm, rather than something independent of all paradigms against which rival interpretations can be measured (1980, 15). Kuhn's paradigm thus has more than a passing resemblance to Stanley Fish's INTERPRET[AT]IVE COMMUNITY, and both concepts have been subjected to similar criticisms.

Lovell suggests that there is a strongly conventionalist element in the work of both Ferdinand de Saussure and Louis Althusser, although they have little else in common. Howard Felperin has argued that there is a conventionalist element in the work of the NEW HISTORICISTS, and he criticizes Stephen Greenblatt in particular for this alleged failing. According to Felperin, Greenblatt's 'theoretical retrenchments of the 1970s and early 1980s' allow him to modify an essentially realist model of history, culture, and literature by the addition of a 'sleeker "textualist" and inevitably "conventionalist" model' (1992, 151; for textualism see the entry for TEXTUALIST).

Thus just as the conventionalist gives up the idea of a reality existing independently of our perceptions of it, and thus makes it impossible to adjudicate between different views of the world in terms of their 'correctness' or conformity to this independently existing reality, the New Historicist loosens his grip on the idea of a historical reality independent of 'narratives' of it and thus drifts in the direction of history as a set of TEXTS which all reflect the beliefs of their creators and between which no adjudication is possible.

See also CONVENTION.

Conversational implicature / maxims See SPEECH ACT THEORY

Co-operative principle See SPEECH ACT THEORY

Copernican revolution A favoured metaphor of much recent theory, used to suggest a range of decentring processes allegedly comparable to the way in which the theories of Copernicus rendered untenable a belief in the earth as the CENTRE of the universe. In his essay 'Freud and Lacan' Louis Althusser refers to Sigmund Freud's comparison (on more than one occasion) of 'the critical reception of his discovery with the upheavals of the Copernican Revolution'. Since Copernicus, Althusser continues,

> we have known that the earth is not the 'centre' of the universe. Since Marx, we have known that the human subject, the economic, political or philosophical ego is not the 'centre' of history – and even, in opposition to the Philosophers of the Enlightenment and to Hegel, that history has no 'centre' but possesses a structure which has no necessary 'centre' except its ideological misrecognition. (1971, 201)

He adds that, in like manner, since Freud we know that the real SUBJECT, 'the individual in his unique essence', has not the form of ego which is centred on consciousness or existence, but rather that 'the human subject is decentred' (1971, 201). Jacques Lacan has an interesting extended discussion of the metaphor of Copernican revolution in his essay 'The Subversion of the Subject and the Dialectic of Desire in the Freudian Unconscious' (1977, 295–6). To this list of 'decenterers' Catherine Belsey has added the name of Ferdinand de Saussure, who, she claims, decentered language and questioned the metaphysics of PRESENCE which had dominated western philosophy (1980, 136).

The relevance of all this for students of literature is that Belsey reaches the conclusion that the epoch of the metaphysics of presence is doomed, 'and with it all the methods of analysis, explanation and interpretation which rest on a single, unquestioned, pre-Copernican *centre*' (1980, 137). In particular, the TEXT (literary or otherwise) is no longer seen as source and centre of its own MEANING; instead, the meaning of the text is detached from a fixed centre and thus deprived of that fixity that comes from self-identity. Such a position ties in with a number of other arguments which have a direct relevance to interpretation: the death of the AUTHOR and the movement from WORK to text.

See also *ÉPISTÉMÈ*; EPISTEMOLOGICAL BREAK; PARADIGM SHIFT; PROBLEMATIC.

Covert plot See STORY AND PLOT

Cratylism From Plato's dialogue *Cratylus*, in which the participants discuss whether names are motivated or not: the theory that the relationship between words and what words refer to is existential rather than CONVENTIONAL.

See also *motivated* and *unmotivated* in the entry for ARBITRARINESS.

Crisis In traditional usage, the point at which the fortunes of the hero change. More recently, Mieke Bal (1985) distinguishes between *crisis* and *development*: the former a short span of time into which many events are compressed, the latter a longer span of time in which, as the name suggests, a development takes place.

Critics of consciousness See PHENOMENOLOGY

Crosstalk According to the linguist John Gumperz, that sort of misunderstanding which is occasioned by having to negotiate complex interactions with, for example, bureaucracies of one sort or another. Thus an Asian living in Britain may have difficulty during a job interview – may get caught in crosstalk – because he or she is abiding by a different set of regulatory CONVENTIONS. (See Gumperz 1982a & 1982b).

See also PUNCTUATION.

Cultural code See CODE

Cultural materialism / cultural poetics See NEW HISTORICISM AND CULTURAL MATERIALISM

Culture According to Raymond Williams's *Keywords*, *culture* is 'one of the two or three most complicated words in the English language' (1976, 76). Williams attributes this complexity partly to the word's intricate historical development in several European languages, but mainly to the fact that it is now used for important concepts in several different intellectual disciplines. His discussion of the word should be consulted in its entirety, but his isolation of three interrelated modern usages is worth summarizing here. These are, first: 'a general process of intellectual, spiritual and aesthetic development'; second: 'a particular way of life', of either a people, a period or a group; third: 'the works and practices of intellectual and especially artistic activity' (1976, 80).

In recent literary-critical discussion the term has undoubtedly been used by those who wish to set literature in a socio-historical context without the use of terms which might invoke a specifically MARXIST methodology or analytical framework, although to this one should add the point that Marxists have offered their own definition of the term. In an article published first in 1937, Edgell Rickword claimed that for the Marxist

> culture is not a mass of works of art, of philosophical ideas, of political concepts accumulated at the top of the social pyramid by specially-gifted individuals, but the inherited solution of problems of vital importance to society. (1978, 103)

This certainly distinguishes culture from, for example, the related terms BASE and SUPERSTRUCTURE, but it also restricts culture to *solutions*. More recent usages, many associated with the newly emergent academic area of Cultural Studies, have seen culture in terms of inheritance, not just of solutions, but of elements from all three of Raymond Williams's alternative definitions. For FEMINISTS it has been important to be able to attribute some or all GENDER characteristics and rôles to cultural rather than to biological influences.

The term *popular culture* refers to the culture of a subordinate group or class which is distinct from the dominant culture of a particular society, dominant in the sense either of more widely disseminated or valued, or in the sense of belonging to and reflecting the interests of a dominant group or class. The term POPULAR is itself problematic, invoking either that which is *for*, or that which is *of*, the people (for which the term *folk culture* has sometimes been reserved). Thus novels by Agatha Christie and Robert Tressell could either or both be included in or excluded from the categories of popular culture or popular fiction depending upon one's definition of these terms.

For *cultural materialism* see NEW HISTORICISM AND CULTURAL MATERIALISM.

See also POPULAR; STRUCTURES OF FEELING.

Cutback See ANALEPSIS

D

Data driven See SOLUTION FROM ABOVE / BELOW

Death of the author See AUTHOR

Declarations See SPEECH ACT THEORY

Decoding See CODE

Deconstruction The term originates in the writings of the French philosopher Jacques Derrida, and implies, as Jonathan Culler puts it, that the hierarchical oppositions of Western metaphysics are themselves constructions or IDEOLOGICAL impositions (1988, 20). Deconstruction thus aims to undermine Western metaphysics by undoing or deconstructing these hierarchical oppositions and by showing their LOGOCENTRIC reliance upon a CENTRE or PRESENCE, which reflects the idealist desire to control the play of signifiers by making them subject to some extra-systemic TRANSCENDENTAL SIGNIFIED. As will be seen, Derrida is a great coiner of neologisms, and to avoid the necessity for repetition the reader is advised to look up these separate terms – along with DIFFÉRANCE, DISSEMINATION, and PHONOCENTRISM. Deconstruction is generally taken to represent an important – even dominant – element in POST-STRUCTURALISM.

There is no absolute agreement concerning what implications Derrida's more general positions hold for literary criticism and theory. At one extreme its implications can appear modest: Jonathan Culler quotes Barbara Johnson to the effect that deconstruction is 'a careful teasing out of warring forces of signification within the text' (Culler 1981, ix) – a statement with which the New Critics would surely have been in full accord. In an interview with Imre Salusinszky Johnson has further commented that

> if it is indeed the case that people approach literature with the desire to learn something about the world, and if it is indeed the case that the literary medium is not transparent, then a study of its non-transparency is crucial in order to deal with the desire one has to know something about the world by reading literature. (Salusinszky 1987, 166)

This, one may be forgiven for noting, is so buttressed with qualifications that it would be hard to disagree with – although it does hedge its bets on whether it is possible to learn something about the world through literature or whether this is a delusion experienced by 'people' who can be relieved of their inappropriate 'desire' through a study of the literary medium's non-transparency.

Johnson does, however, go on to distance herself and deconstruction from the 'self-involved textual practice of "close reading"' of the New Critics mentioned by her interviewer, suggesting that deconstruction necessarily involves a political attitude, one which examines authority in language, and she notes that Karl Marx was as close to deconstruction as are a lot of deconstructors – particularly by virtue of his bringing to the surface the hidden inscriptions of the economic system, uncovering hidden presuppositions, and showing contradictions (Salusinszky 1987, 167).

It would certainly seem that deconstruction involves one inescapable implication for any process of interpretation. This is that the interpretation of a TEXT can never arrive at a final and complete 'meaning' for a text. As Derrida himself remarks about a READING of the Marxist 'classics',

> These texts are not to be read according to a hermeneutical or exegetical method which would seek out a finished signified beneath a textual surface. Reading is transformational. (1981b, 63)

Not just reading, but (clearly implied) each reading. Thus for Derrida the MEANING of a text is always unfolding just ahead of the interpreter, unrolling in front of him or her like a never-ending carpet whose final edge never reveals itself. Introducing a volume of essays entitled *Post-structuralist Readings of English Poetry*, the volume's editors, Richard Machin and Christopher Norris note that post-structuralist readings tend to 'feature the text as active object' (1987, 3): the AUTHOR is no longer seen as the source of meaning, and deconstruction is guilty of being an accessory after the fact with regard to the death of the author. Later on in their introduction they seek to establish that whereas each reading in the collection 'develops an insistent coherence of its own that drives towards conclusive and irrefutable conclusions', the possibility is nonetheless held open of 'a multitude of competing meanings, each of which denies the primacy of the others' (1987, 7). The possibility of such a paradoxical blending of linear rigour and pluralistic co-existence has not always convinced the sceptical, however, and one of the most recurrent criticisms of the readings or interpretations generated by deconstruction is that they are not subject to falsification. Another objection is that these readings and interpretations have a tendency to end up all looking the same, all demonstrating the ceaseless play of the signifier and nothing much else, just as crude psychoanalytic readings of the 1930s and 1940s tended all to end up demonstrating certain recurrent items of Freudian faith. And indeed the more a criticism holds that interpretations are not subject to the control of textual meaning (however defined), the more it has to cope with the problem that the choice of text necessarily becomes a matter of less and less moment. How can one talk about rigorously grappling with a text if there is said to be nothing fixed 'in' the text?

See also HERMENEUTICS.

Deep structure See STRUCTURE

Defamiliarization Also singularization. From the Russian meaning 'to make strange', the term originates with the RUSSIAN FORMALISTS and, in particular, the theories of Viktor Shklovsky. In his essay 'Art as Technique', Shklovsky argues that perception becomes automatic once it has become habitual, and that the function of art is to challenge automization and habitualization, and return a direct grasp on things to the individual perception.

> Habitualization devours works, clothes, furniture, one's wife, and the fear of war. 'If the whole complex lives of many people go on unconsciously, then such lives are as if they had never been.' And art exists that one may recover the sensation of life; it exists to make one feel things, to make the stone *stony*. (1965, 12; the quotation is from Leo Tolstoy's *Diary*)

The PRAGUE SCHOOL theorist Bohuslav Havránek provides useful definitions of both *automization* and *foregrounding* in his essay 'The Functional Differentiation of the Standard Language'.

> By *automization* we . . . mean . . . a use of the devices of the language, in isolation or in combination with each other, as is usual for a certain expressive purpose, that is, such a use that the expression itself does not attract any attention . . .
>
> By *foregrounding* . . . we mean the use of the devices of the language in such a way that this use itself attracts attention and is perceived as uncommon, as deprived of automization, as deautomized, such as a live poetic metaphor (as opposed to a lexicalized one, which is automized). (1964, 9, 10)

For Shklovsky the purpose of art is to impart the sensation of things as they are perceived and not as they are known, and the technique used to achieve this end is that of making objects *unfamiliar*. One variant of this technique which Shklovsky attributes to Tolstoy is that of not naming an object but of describing it as if one were seeing it for the first time (1965, 13). This technique actually predates Tolstoy by many years: think for example of the Lilliputian descriptions of the contents of Gulliver's pockets in Jonathan Swift's *Gulliver's Travels*. Boris Tomashevsky, one of Shklovsky's fellow Russian Formalists, refers to Swift's defamiliarizing techniques in *Gulliver's Travels* in his essay 'Thematics' (1965, 86).

It is important to note that members of the Prague School believed that automization could occur at various levels: a CANON, for instance, could become automized, and subsequently defamiliarized by a work which forced readers to recognize this fact.

The term *foregrounding* represents a development of the concept of defamiliarization which builds on FIGURE-GROUND theories developed by researchers into perception. Nowadays, *foregrounding* and *defamiliarization* are often used interchangeably. Both concepts are related to the view that 'poetic language' is to be sharply distinguished from other forms of language; according to Jan Mukařovský, the 'function of poetic language consists in the maximum of foregrounding of the utterance', while the 'foregrounding of any one of the components is necessarily accompanied by the automatization of one or more of the other components' (1964, 19–20). *Foregrounding* has led to the coining of a complementary term: *backgrounding* – meaning the process whereby certain elements in a literary WORK are presented in such a way as *not* to stand out or be noticed. The most overt example of backgrounding is probably to be found in detective novels, in which part of the AUTHOR's skill lies in blending in crucial clues with the unperceived background of the work so that the reader only fully understands the full significance of the clue later on. (See also the entry for ELLIPSIS for Gérard Genette's comments on *hypothetical ellipses*; and also the entry for DEFERRED / POSTPONED SIGNIFICANCE.)

The verb *to naturalize*, which also gives us the process of *naturalization*, is often used as an alternative term in English for automization. According to Gérard Genette, for Roland Barthes the 'major sin of petty-bourgeois ideology' is the naturalization of culture and history (1982, 36). By this is meant that CULTURE and history are made so familiar that their historico-cultural specificity, and thus the possibility of changing them, is obscured. Anne Cranny-Francis has argued that part of the formation of subjectivity of men and women involves the naturalization of SEXIST DISCOURSE 'as the obvious mode of representation and self-representation of women and men' (1990, 2).

See also DEFORMATION; LITERARINESS.

Default interpretation According to Geoffrey Leech, the 'initial and most likely interpretation' that will be accepted '*in default* of any evidence to the contrary' (1983, 42). The term comes from Leech's book *Principles of Pragmatics* and is proposed in the context of a discussion about the PRAGMATICS of ordinary interpersonal speech communication. So far as literature is concerned it will be clear that what one's initial and most likely interpretation of a TEXT or a part of a text is will be dependent upon a range of contextual factors, including those relating to genre. Now although contextual factors also influence and condition interpersonal speech communication, the fact that an UTTERANCE is delivered in person to its recipient means that the extreme variation of contextual factors with which one can meet when interpreting a literary text is unlikely to be encountered. For this reason, the concept of default interpretation is likely to be of limited value in literary studies.

Deferred / postponed significance Also referred to as *enigma* by Roland Barthes. Any element in a NARRATIVE the full significance of which is only appreciated at a stage later in the telling than that at which it appears.

Compare *delayed decoding* in the entry for CODE.

Deformation In Roman Jakobson's essay 'On Realism in Art' the term *deformation* is given a meaning very similar to that given to the term DEFAMILIARIZATION by the RUSSIAN FORMALISTS. According to Jakobson, as artistic traditions develop, 'the painted image becomes an ideogram, a formula, to which the object portrayed is linked by contiguity' (1971, 39). (Compare what is said about *automization* in the entry for defamiliarization.) To break into this STEREOTYPED encoding of reality in art, according to Jakobson, the ideogram 'needs to be deformed' (1971, 40), and he outlines a number of different ways in which this can be done. He thus makes REALISM a RELATIVE term, dependent upon the operation of processes of disruption which prevent artistic CONVENTIONS from hardening into what he calls ideograms, or stereotypes.

For the PRAGUE SCHOOL theorists deformation is linked to the concept of *foregrounding* (see the entry for defamiliarization). According to them, deformation can be used either to make an element stand out so that it is foregrounded, or (as Yuri Tynyanov in particular argues), in the opposite manner: so as to reduce non-foregrounded elements to the status of neutral props.

See also DEVIATION; DOMINANT.

Degree zero (writing) See ÉCRITURE

Deixis Those features of language which fasten utterances temporally or spatially: 'here', 'now', for example. *Deictics*, or *deictic elements* (*deictic* can perform as either noun or adjective) play an important rôle in NARRATIVE; they constitute an important token of FREE INDIRECT DISCOURSE, for example.

See also SHIFTER.

Delayed decoding See CODE

Denaturalize See NATURALIZE

Denotation See CONNOTATION AND DENOTATION

Desire Both as noun and as verb *desire* indicates a central but diffuse and by no means unified concept or set of concepts in a cluster of different contemporary theories, very often in connection with attempts to DECONSTRUCT or theorize the SUBJECT or subjectivity. According to Michel Foucault, the more recent researches of psychoanalysis, linguistics and ethnology have 'decentred the subject in relation to', among other things, 'the laws of his desire' (1972, 13), and the concept of desire has assumed an important but varied function within theories concerned to see the subject as more SITE than determining origin or PRESENCE. For Jacques Lacan, because the subject is split between a conscious mind the contents of which are unproblematically retrievable, and an unconscious set of drives and forces (*Trieb*), and because the subject knows that what it knows is not all that it is, desire for the *other* is a constituting part of the subject. This is my précis of a notoriously difficult author's position; those wishing to check it should consult, among other sources, Lacan 1977, 292–325. The following quotation illustrates the difficulty involved:

> For it is clear that the state of nescience in which man remains in relation to his desire is not so much a nescience of what he demands, which may after all be circumscribed, as a nescience as to where he desires.
>
> This is what I mean by my formula that the unconscious is '*discours de l'Autre* (discourse of the Other), in which the *de* is to be understood in the sense of the Latin *de* (objective determination): *de Alio in oratione* (completed by: *tua res agitur*).
>
> But we must also add that man's desire is the *désir de l'Autre* (the desire of the Other) in which the *de* provides what grammarians call the 'subjective determination', namely that it is *qua* Other that he desires (which is what provides the true compass of human passion). (1977, 312)

For Lacan, moreover, desire is necessarily linked to PHALLOCENTRISM because the child desires the mother's desire and thus identifies himself (Lacan's gendered term) 'with the imaginary object of this desire in so far as the mother herself symbolizes it in the phallus' (1977, 198). Desire is also, according to Lacan, constituted by the hysteric in the very moment of speaking. 'So it is hardly surprising that it should be through this door that Freud entered what was, in reality, the relations of desire to language and discovered the mechanisms of the unconscious' (1979, 12).

Vincent Crapanzano has provided an accessible account of the distinction Lacan makes between desire and need:

> Desire, Lacan has written, is 'an effect in the subject of that condition which is imposed upon him through the defiles of the signifier' (Wilden 1968, 185). Need is directed toward a specific object, is unmediated by language, and, unlike desire, can be satisfied directly. (Desire must always be satisfied, insofar as it can be satisfied, by symbolic substitutes for that which it can never possess [Crapanzano 1978].) To become a self, the individual must seek recognition by demanding the other to recognize him-self, or his desire – to acknowledge at least the noun (his name) as a (grammatically) legitimate *Anlage* for the I and the you. The individual must take possession of his own otherness and not be aware simply of the otherness about him. (1992, 89)

FEMINIST critics, not surprisingly, have displayed both interest in and suspicion towards the concept of desire, variously defined. Catharine A. MacKinnon, for example, states that she has selected 'desire' as a term parallel to 'value' in MARXIST theory, 'to refer to that substance felt to be primordial or aboriginal but posited by the theory as social and contingent' (1982, 2). MacKinnon distances herself forcefully from the use of the term 'desire' to be found both in Jean-Paul Sartre's *Existential Psychoanalysis* and in *Anti-Oedipus: Capitalism and Schizophrenia* by Gilles Deleuze and Felix Guattari. In these works, MacKinnon argues, the concept of desire entails sexual objectification, which for her is 'the primary process of the subjection of women' (1982, 27). To substantiate her case she quotes first Sartre: 'But if I desire a house, or a glass of water, or a woman's body, how could this glass, this piece of property reside in my desire and how can my desire be anything but the consciousness of these objects as desirable?' (Sartre 1973, 20), and then Deleuze and Guattari's view of man as 'desiring-machine'. She insists: 'Women are not desiring-machines' (1982, 27).

See also BODY; SYNTAGMATIC AND PARADIGMATIC (metonymy as desire).

Destinataire / destinateure See ACT / ACTOR

Determinate absence See ABSENCE

Determinate instance See INSTANCE

Deviation The more stress is laid upon NORMS and CONVENTIONS in a given theory or approach, the more significance is likely to be accorded to deviation from these norms and conventions. (We can only confidently refer to 'deviates' in a CULTURE in which we have, or think we have, a firm sense of what constitutes normality.) Clearly deviation is closely related to the RUSSIAN FORMALIST concept of DEFAMILIARIZATION, in which the LITERARINESS of language consists in the extent to which it deviates from extra-literary language, or to which it encourages deviation from everyday habits of perception. In the context of the Russian Formalists, then, deviation and DEFORMATION are well-nigh interchangeable as terms.

Deviation is also closely related to various usages of the terms DIFFERENCE and DIFFÉRANCE, as a deviation is significant as much (if not more) in terms of what it is *not* as in terms of what it *is*. Deviation has also become an important term in the fields of STYLISTICS and NARRATOLOGY: a style may be constituted at least in part by deviations from a linguistic norm, and according to Gérard Genette, Marcel Proust's *À la Recherche du Temps Perdu* deviates from then-accepted laws of NARRATIVE by its manipulation of the *singulative* and the *iterative modes* – basing the narrative rhythm of this work not, as in the classical novel, on alternation between the summary and the scene, but on alternation between the iterative and the singulative mode (1980, 143). (For an explanation of these terms see the entry for FREQUENCY.)

Device See FUNCTION

Diachronic and synchronic A diachronic study or analysis concerns itself with the evolvement and change over time of that which is studied: thus diachronic linguistics is also known as historical linguistics, and is concerned with the development of a language or languages over time. A synchronic study or analysis, in contrast, limits its concern to a particular moment of time. Thus Synchronic Linguistics takes a language as a working system at a particular point in time without concern for how it has developed to its present state. One of the main reasons why Ferdinand de Saussure is credited with having revolutionized the study of language early in this century is that he drew attention to the possibility of studying language synchronically, and thus established the possibility of a STRUCTURAL linguistics. There are strong grounds for attributing a belief in the necessity of both the diachronic and the synchronic study of language to Saussure; at the start of the second chapter of the *Course*, outlining what the scope of linguistics should be, he includes the task of describing and tracing the history of all observable languages (1974, 6), and later on he suggests that

> the thing that keeps language from being a simple convention that can be modified at the whim of interested parties is not its social nature; it is rather the action of time combined with the social force. If time is left out, the linguistic facts are incomplete and no conclusion is possible. (Saussure 1974, 78)

(It is, additionally, worth mentioning that this comment also tends to the invalidation of a common 'Saussurean' myth – that the ARBITRARINESS of language means that words can mean anything one wants them to mean.)

Most conclusively, in the *Course* Saussure speaks directly of the reasons for distinguishing 'two sciences of language', *evolutionary linguistics* and *static linguistics* (1974, 81).

The extent to which synchronic study really does as it were take a frozen slice of history for study is itself not absolute: to talk of a SYSTEM is necessarily to imply movement and interaction, and movement and interaction can only take place in time. Thus the synchronic studies of complete CULTURES carried out by the anthropologist Claude Lévi-Strauss involved investigation of, for instance, symbolic exchanges which were consecutive rather than simultaneous, so that the element of temporal sequence is still present in such STRUCTURALIST investigations.

There have been attempts to study literature synchronically (diachronic literary study has, of course, a long pedigree). Roman Jakobson, for example, has argued that the synchronic description of literature concerns itself not just with present-day literary production, but also with that part of the literary tradition which has either remained vital or has been revived (1960, 352). Gérard Genette, taking up this suggestion from Jakobson, suggests that the structural history of literature 'is simply the placing in diachronic perspective of these successive synchronic tables' (1982, 21). This of course leaves unanswered (or unacknowledged) the question of whether there are isolable laws of transformation governing the replacement of one synchronic table by its successor, as to whether, in other words, historical development itself is capable of being analysed in terms of cause and effect. (There is also the question of the interaction between different, co-temporal structures, which may only be observable historically.)

The concept of 'structural history', indeed, implies that the fundamental reality is that of the synchronic structure, and that history is a secondary reality formed by successive structures.

Diacritical Originally an adjective which conveyed the idea of distinguishing or conferring distinction: thus diacritical marks are those signs added to letters to give them distinction – the cedilla which changes c to ç for instance. By process of association the term has assumed a noun form within the field of Linguistics to signify diacritical marks.

Within STRUCTURALIST usage *diacritical* is granted a slightly different, more general adjectival meaning: to say that language is a diacritical system is to claim that it is a system constituted by significant DIFFERENCES.

Dialect See IDIOLECT

Dialectics From a Greek word meaning debate or argument, dialectics originally referred to the process of revealing the truth by argument or debate – especially when this involved revealing contradictions in one's opponent's arguments. More recently the term has been used to describe (i) a philosophical outlook which considers all things to exist in dynamic relationships and to be possessed of internal tensions and contradictions, and (ii) a method of investigating reality which stresses the dynamic interconnections of things in the world and of their internal tensions and contradictions.

Dialogic Along with *dialogue*, *dialogism*, *dialogical*, and *dialogism*, dialogic owes its current technical use to the influence that the writings of Mikhail Bakhtin have had in the West following their translation into English during the 1970s and 1980s. Bakhtin wrote under difficult conditions in the Soviet Union of the 1920s and, especially, the 1930s (his first published work was in 1919 and he was still writing at the time of his death in 1975). As a result of these difficult circumstances it has been claimed (and disputed) that some of his writings were published under the names of his friends V. N. Vološinov and P. N. Medvedev. In what follows I shall attribute opinions to the person whose name was associated with them on initial publication.

Dialogue in its everyday usage means verbal interchange between individuals, especially as represented in literary writing. Vološinov (1986) builds upon this familiar usage in a number of ways. First, he suggests that verbal *interaction* is the fundamental reality of language: both in the history of the individual and also in the history of the human species, language is born not within the isolated human being, but in the interaction between two or more human beings. The recent development of PRAGMATICS – both in Linguistics and Literary Studies – may make this obvious to the present-day reader, but highly influential theories of language have obscured this truth. Second, however, Bakhtin was increasingly to argue that even in DISCOURSE or UTTERANCE which was not overtly interactive, dialogue was to be found. Because all utterances involve the, as it were, 'importing' and naturalization of the speech of others, all utterances include inner tensions, collaborations, negotiations which are comparable to the process of dialogue (in its everyday sense). For Bakhtin, words were not neutral; apart from neologisms (of which he was, not surprisingly, rather fond) they were all second-hand and had belonged to other people, and in incorporating them into his or her own usage the individual had to engage in dialogue with that other person, struggle to wrest possession of them from their previous owner(s). Discussing language, Bakhtin habitually makes use of terms such as 'saturated'; 'contaminated'; 'impregnated'; a word for Bakhtin is like a garment passed from individual to individual which cannot have the smell of previous owners washed out of it. Spoken or written utterances are like palimpsests: scratch them a little and hidden meanings come to light, meanings which are very often at odds with those apparent on the surface.

The Bakhtinian view of the dialogic connects with the topics of INTERTEXTUALITY and transtextuality, and with Harold Bloom's concept of the anxiety of influence (see the entry for REVISIONISM). For in using a word or an expression, an author will engage in some sort of dialogue with the text in which he or she first encountered this word, or the text in which the word has had a particular meaning embossed upon it.

The opposite of dialogue for Bakhtin is, logically, *monologue*.

> Ultimately, *monologism* denies that there exists outside of it another consciousness, with the same rights, and capable of responding on an equal footing, another and equal *I (thou)*. . . . The monologue is accomplished and deaf to the other's response; it does not await it and does not grant it any *decisive* force. (Quoted in Todorov 1984, 107)

Polyglossia is Bakhtin's term for the simultaneous existence of two national languages within a single CULTURAL system; in contrast, *monoglossia* indicates that a culture contains but one national language.

Interest in Bakhtin has sent commentators back to the work of the German Hans-Georg Gadamer, in whose major work *Truth and Method* the concept of dialogue also looms large. Gadamer distinguishes between authentic and inauthentic conversation and suggests that a reader's encounter with a text is like authentic conversation – open, two-sided, unegocentric (1989, 385).

See also INTERIOR DIALOGUE.

Diegesis and mimesis Both of these terms are to be found in the third book of Plato's *Republic*, in which Socrates uses them to distinguish between two ways of presenting speech. For Socrates, diegesis stands for those cases where the poet himself is the speaker and does not wish to suggest otherwise, and mimesis stands for those cases in which the poet attempts to create the illusion that it is not he who is speaking. Thus a speech spoken by a CHARACTER in the play would represent mimesis, whereas if the writer spoke 'as him or herself' about characters, we would have a case of diegesis. Monika Fludernik has stressed that these terms must be understood in relation to the artistic genres and practices from which they emerged and to which they were designed to apply.

> When Plato opposes narration (diegesis) to mimesis (imitation), he does so within a generic contrast of drama (all imitation) versus the dithyramb (all diegesis) and the epic (mixed), and the diegesis is here not simply 'narrative', but very explicitly the author's direct voice. For Plato, of course, does not distinguish between a narrator and the author, which is the only way that the narrator of the epic would become identified with the speaker of elegiac verse. In narrative theory Plato's distinctions have been forgotten, with the result that diegesis has come to signify narration, *per se*, that is to say the *narrator's* rendering of the story in everything except clearly defined discourse not attributable to this enunciator, i.e. the characters' directly quoted speech and thought acts. This schema makes it very difficult to conceptualize action (or, for that matter, description). (1993, 28)

Fludernik illustrates this distinction by means of a diagram taken from Irene de Jogn (1989):

	diegesis haplè	*mimesis*
the poet speaks as	himself	one of the characters
in the	parts between the speeches	speeches

She explains that

> Narrative is here considered 'pure' or 'simple' (*haplè*), i.e. 'single-voiced' speaking, whereas in mimetic narration the narrator impersonates one of the characters, and the discourse therefore becomes double-voiced, dual or double-levelled. (1993, 30)

Aristotle extended use of the term mimesis in his *Poetics* to include not just speech but also imitative actions, and as these could of course be rendered in indirect speech this extension had the effect of blunting Plato's rather sharper distinction.

In his Appendix to Aristotle's *Poetics*, D. W. Lucas (1968) notes that it is clear that not just poetry, painting, sculpture and music are forms of mimesis for both Plato and Aristotle, but so too is dancing. Lucas further adds that the word mimesis has an extraordinary breadth of meaning which makes it difficult to discover just what the Greeks had in mind when they used it to describe what poet and artist do, and he suggests that to translate it we may at different times need to use words such as 'imitate', 'indicate', 'suggest' and 'express', although all of these words are related to human action ('praxis').

Since Aristotle, both terms have been incorporated into different systems of terminology, and their original, clear-cut meanings have been extended into new usages and more complex (and often confusing) distinctions. The term 'mimesis' has been pressed into service to describe the more general capacity of literature to imitate reality, and has on occasions accumulated a somewhat polemical edge as a result of its use by those wishing to establish imitation as central or essential to art – by MARXIST critics intent on stressing that literature and art 'reflect' extra-literary reality, for example.

With the growth of modern NARRATIVE theory in recent years, diegesis in particular has been given something of a new lease of life. A number of theorists have equated diegesis and mimesis with telling and showing, a distinction which can be traced back to Henry James, and which was both adopted and simplified by Percy Lubbock. This actually makes rather a large difference in the meaning of diegesis. For if one takes a novel such as, for example, Jane Austen's *Pride and Prejudice*, what the READER learns he or she learns through a *telling*, a *narration*, rather than as a result of a *performance* as in a play. One could thus defend referring to the work *in toto* as an example of diegesis, because even the direct speech of characters is *told* to the reader through a narrating. But any reader of the novel will recognize that such passages in the novel, in which dialogue and character interaction give a dramatic effect, cause the reader to forget about the NARRATOR and to feel as if he or she is witnessing the characters in dramatic interaction. (Note that we have now moved from a concern with the author to a concern with the narrator.) From a Jamesian perspective such passages would be categorized as examples of showing rather than telling, so that if diegesis and mimesis are to be treated as equivalent to telling and showing, then clearly *Pride and Prejudice* is not an example of pure diegesis, but includes both diegetic and mimetic elements. Gérard Genette comments:

> no narrative can 'show' or 'imitate' the story it tells. All it can do is tell it in a manner which is detailed, precise, 'alive,' and in that way give more or less the *illusion of mimesis* – which is the only narrative mimesis, for this single and sufficient reason: that narration, oral and written, is a fact of language, and language signifies without imitation. (1980, 164)

Modern narrative theory has introduced another use of these terms which is related to those discussed already, but which actually represents a significant change and extension of their meaning. Instead of relating them to telling and showing, it has equated them with PLOT and STORY, such that the diegetic level is the level of the 'story reality' of the events narrated, while the mimetic level is the level of the 'narrator's life and consciousness'. Both Genette and Rimmon-Kenan, for example, use *diegesis* as 'roughly equivalent to my "story"' (Rimmon-Kenan 1983, 47). This extension can lead us into some terminological contradictions. For if diegesis is equivalent to story, then *extra-diegetic* must mean 'outside the story', and therefore could refer us to the actual *telling* of the story, the comments from a narrator who is not a member of the world of the story. But this is exactly the opposite of what we started with: for Socrates, we may remember, diegesis referred to those cases where the poet himself is the speaker, roughly what we have just termed *extra-diegetic*! In narrative theorists such as Shlomith Rimmon-Kenan and Gérard

Genette then, an extradiegetic narrator is a narrator who, like the narrator of *Pride and Prejudice*, exists on a different narrative level from the level of the events narrated or the story, whilst an intradiegetic narrator is one who is presented as existing on the same level of reality as the characters in the story he or she tells: Esther Summerson in Charles Dickens's *Bleak House*, for example. Non-personified narrators introduce an additional problem here, but it is still possible to refer to 'extra-diegetic narrative' even in these cases, for the narrative 'knows' things which the characters do not and could not.

To add to the confusion (as he himself admits), in his *Narrative Discourse* Gérard Genette uses the term *metadiegetic* to describe 'the universe of the second narrative', and the term *metanarrative* to refer to 'a narrative within the narrative' (1980, 228n). This is confusing because a METALANGUAGE is a language about a language – in other words, a 'framing' language and not a 'framed' one. For this reason Genette's terminology has not caught on (as can be confirmed by reference to François Lyotard's definition of the postmodern as 'incredulity towards metanarratives', by which he means narratives *about* and not *within* other narratives: see the entry for MODERNISM AND POSTMODERNISM). Genette's own usage is inconsistent, and in essays collected in his *Figures of Literary Discourse* he uses a different terminology, defining a *metalanguage* as a 'discourse upon a discourse' (he defines criticism as a metalanguage) and a metaliterature as 'a literature of which literature itself is the imposed object' (1982, 3–4). Rimmon-Kenan's alternative term for the level of the EMBEDDED narrative – *hypodiegetic* – avoids these confusions (1983, 92).

A number of recent narrative theorists have suggested that, rather than considering mimesis and diegesis as two mutually exclusive categories, it is more productive to think of them as representing a continuum with minimal narrator colouring at one end and maximal narrator colouring at the other.

The terms *homodiegetic* and *heterodiegetic*, coined by Gérard Genette, introduce additional complications. Genette applies both terms to distinguish different types of ANALEPSIS or flashback: whereas a homodiegetic analepsis provides information about the same character, or sequence of events or milieu that has been the concern of the text up to this point, a heterodiegetic analepsis refers back to some character, sequence of events, or milieu *different from* that/those that have been the concern of the preceding text.

Différance A portmanteau term coined by Jacques Derrida, bringing together (in its French original) the senses of DIFFERENCE and deferment. For Derrida différance is the opposite of and alternative to LOGOCENTRISM; while logocentrism posits the existence of fixed MEANINGS guaranteed by an extra-systemic PRESENCE or origin, différance sees meaning as permanently deferred, always subject to and produced by its difference from other meanings and thus volatile and unstable. Meaning is always relational, never self-present or self-constituted. Derrida uses and discusses the term throughout his writing, but perhaps the most accessible of his discussions is in *Positions* (1981b, 26ff), in which he identifies three main meanings for the term:

> *First*, *différance* refers to the (active *and* passive) movement that consists in deferring by means of delay, delegation, reprieve, referral, detour, postponement, reserving. . . . *Second*, the movement of *différance*, as that which produces different things, that which differentiates, is the common root of all oppositional concepts that mark our language, such as, to take only a few examples, sensible/intelligent, intuition/signification, nature/culture, etc. . . . *Third*, *différance* is also the production, if it can still be put this way, of these differences, of the diacriticity that the linguistics generated by Saussure, and all the structural sciences modeled upon it, have recalled is the condition for any signification and any structure. . . . From this point of view, the concept of *différance* is neither simply structuralist, nor simply geneticist, such an alternative itself being an 'effect' of *différance*. (1981b, 8–9)

Derrida has suggested a number of alternative terms for différance, including (in *Positions*) *gram*.

It should be apparent from what has been said that the attempt to provide a neat, glossary definition of this term raises certain logical problems, for if one accepts Derrida's argument then the meaning of différance, like that of any other term, is deferred and subject to difference: there is no firm or fixed presence that can guarantee or underwrite the meaning of the term. If there were, then the theory on which the term depends would be in error.

Difference In an essay on the varied meanings attributed to this term within FEMINIST theory alone, Michèle Barrett expresses surprise at 'what can be fitted into this capacious hold-all of a concept', and admits to a lack of clarity concerning the term's meaning within different contexts (1989, 38). If we seek to explain the rise to prominence of the term in the last couple of decades then the best place to start is with Ferdinand de Saussure's *Course in General Linguistics*. Absolutely fundamental to Saussure's approach is the view that language works as a system of differences – that, as he puts it, 'in a language-state everything is based on relations' (1974, 122). He points out, for example, that the modern French *mouton* and the English *sheep* have the same signification but not the same value, because the single French word is roughly equivalent to *two* English ones: *sheep* and *mutton*. Thus the value of *sheep* is partly conditioned by its being *not-mutton* – by being *different* from *mutton*.

Saussure adds that 'everything said about words applies to any term of language, e.g. to grammatical entities' (1974, 116). It is certainly true of the phonemic system, where it is not necessary that all speakers of a language produce identically sounding phonemes, but that the same set of phonemically significant *differences* between sounds can be recognized. On the syntactical level, Saussure draws attention to the importance of SYNTAGMATIC and PARADIGMATIC choices, thus providing two key axes of meaning-generating difference in the syntax of the sentence.

Given the significance of Saussure's work for the development of important theoretical movements such as STRUCTURALISM, the idea that it is difference rather than identity or PRESENCE that is important has won widespread acceptance in a range of theoretical fields. Jacques Derrida's coinage DIFFÉRANCE is a development of a Saussurean theme – even though one which deconstructs its own parent. The idea that where signification is concerned, what something *is* is dependent upon what it *is not*, that meaning is generated at least in part by a difference from what is *not meant*, has clear points of contact with theories of determinate ABSENCE in the interpretation of complex TEXTS – including literary ones. MEANING in literary and CULTURAL texts is in part (those who believe that systems of signification are closed would say wholly) generated by paradigmatic exclusions, by a difference from the not-meant. A smart three-piece suit has significance at least in part because of what it is not, as a result of its displayed difference from other possible styles of dress – just as sheep is sheep partly by virtue of being not-mutton.

Jonathan Culler suggests one way in which some of these insights can be applied to literature:

> If in language there are only differences with no positive terms, it is in literature that we have least cause to arrest the play of differences by calling upon a determinate communicative intention to serve as the truth or origin of the sign. We say instead that a poem can mean many things. (1975, 133)

Michèle Barrett's essay, mentioned at the start of this entry, provides some representative examples of the appropriation and development of this term within feminist theory. Barrett isolates three main feminist usages of the concept of difference: (i) the 'sexual difference' position, seen for example in psychoanalytic discussion which invokes an ESSENTIALIST conception of GENDER identity and gendered subjectivity, and which explicitly refuses the sex/gender distinction; (ii) a more Saussurean view of meaning as positional or relational which, in the form of a POST-STRUCTURALIST and anti-MARXIST critique of totality, sees GENDER, race and class as SITES of difference (rather than, as Barrett points out, sites of the operation of power), and which DECONSTRUCTS the idea of gendered subjectivity; and (iii) a more diffuse usage which stresses plurality and diversity within, say, feminism. Barrett suggests that these different usages may not be reconcilable, and that the use of the same word to cover all three may therefore be unwise.

Digital and analogic communication The distinction between digital and analogic communication is illustrated by Paul Watzlawick *et al.* through reference to the difference between the basic modes of operation of the central nervous system and of the humoral system. In the central nervous system

> the functional units (neurons) receive so-called quantal packages of information through connecting elements (synapses). Upon arrival at the synapses these 'packages' produce excitatory or inhibitory postsynaptic potentials that are summed up by the neuron and either cause or inhibit its firing. (1968, 60)

The central nervous system works by means of the DIGITALIZATION of information via a mass of BINARY possibilities: a neuron either fires or it does not: there are no shades of grey. The humoral system, in contrast, works in a completely different manner: by means of the release of discrete quantities of particular substances into the bloodstream. Releasing either more or less of a given substance can lead to a significantly different effect. Watzlawick *et al.* note that these two systems 'complement and are contingent upon each other'.

A perhaps more familiar example of the same distinction would be that between the two sorts of loudness indicator on a tape recorder. The traditional moving-needle indicator is an analogue system: each slight variation in noise produces a variation in the position of the needle. In contrast, the LED system consists of a series of lights which either fire or they do not. Modern computer systems are based upon the digitalization of information.

Theories based upon the digitalization of information have assumed a considerable importance in Linguistics in recent years. In phonetic systems, for example, research may reveal that different members of a language community may actually produce rather different sounds to represent a particular phonetic units, but so long as these sounds produce recognizable *oppositions* or binary distinctions, the system works. The Linguistics of Ferdinand de Saussure relies heavily upon the recognition of DIFFERENCES between binary oppositions.

All this would seem to be a long way away from literature and literary criticism. But Jonathan Culler has pointed out that the linguistic model has encouraged STRUCTURALISTS to think in binary terms and to search for functional oppositions in whatever material they are studying (1975, 14), and this has had a direct influence upon such fields of knowledge as NARRATIVE and literary criticism.

It has, for example, created an awareness that the READER's perception of a set of binary oppositions may play a necessary function in the READING of a WORK of literature, alongside the operation of more subtle, 'analogic' responses. This is true at the macro-level of genre and generic classification (for example, the opposition between tragedy and comedy sets up certain either-or expectations which exert a strong influence on our literary responses), and also at micro-levels within particular TEXTS. An example here would be Claude Bremond's narrative theory in which each function opens for two possibilities: ACTUALIZATION or non-actualization.

Certainly in FORMULAIC literature response seems to be conditioned by the operation of simple binary distinctions: 'If this blonde, shy girl is the heroine, then this dark, self-confident girl must be the attractive but treacherous rival.'

Directives See SPEECH ACT THEORY

Disavowal A term from Freudian psychoanalysis referring to the way in which a person under analysis will simultaneously affirm and deny something – or will affirm something *by* denying it.

See also REPRESSION.

Discourse This word has experienced a relatively sudden rush of fashionability in the past couple of decades in a number of different academic and intellectual fields. According to the OED, *discourse* as noun can mean (sense 4) 'Communication of thought by speech', and Samuel Johnson's definition is quoted: 'Mutual intercourse of language.' Interestingly, the use of the noun to mean 'talk' or 'conversation' is described as archaic.

In Linguistics a renewed reliance upon the term is related to the growth in importance of PRAGMATICS; discourse is language in use, not language as an abstract system. But even within Linguistics there are varieties of meaning. Michael Stubbs comments on the use of the terms TEXT and *discourse*, and states that this is often ambiguous and confusing. He suggests that the latter term often implies greater length than does the former, and that *discourse* may or may not imply interaction (1983, 9). Thus if we take an academic seminar, for some linguisticians the whole process of verbal interaction would constitute a discourse, whereas for others an extended statement by one participant would qualify as a discourse. Yet others would be prepared to accept even short statements by individuals as discourses. Moreover, for some linguisticians discourse is uncountable, for others it is not, and for yet others it appears to be countable at some times but not at others. If discourse *is* countable, the next problem is to decide what constitute(s) the defining borders of a single discourse: Michael Stubbs notes that the unity of a particular discourse can be defined in either structural, semantic or functional ways (1983, 9).

Gerald Prince isolates two main meanings for the term within NARRATIVE theory: first, the expression plane of a narrative rather than its content plane, the narrating rather than the narrated. Second, following Benveniste, *discourse* is distinguished from *story* (*discours* and *histoire* in the original French) because the former evokes a link between 'a state or event and the situation in which that state or event is linguistically evoked' (Prince 1988, 21). Contrast 'John's wife was dead' (story) with 'He told her that John's wife was dead' (discourse). (Compare the distinction between *énonciation* and *énoncé* in the entry for ENUNCIATION.) Some writers on narrative in English prefer to retain *discours* in untranslated form when using the term in Benveniste's sense.

The work of Michel Foucault has been highly influential across a number of disciplines so far as the term discourse is concerned. For Foucault discourses are 'large groups of statements' – rule-governed language terrains defined by what Foucault refers to as 'strategic possibilities' (1972, 37), comparable to a limited extent to one possible usage of the term REGISTER in Linguistics. Thus for Foucault at a given moment in the history of, say, France, there will be a particular discourse of medicine: a set of rules and CONVENTIONS and SYSTEMS of MEDIATION and transposition which govern the way illness and treatment are talked about, when, where, and by whom. Clearly we meet a similar problem here to that mentioned in a different context above: how does one define the boundaries of a particular discourse?

Foucault also uses the term *discursive formation* in a way that seems roughly interchangeable with *discourse*: *discursive* here represents the adjective form of discourse, not the adjective meaning 'round-about, meandering'. (John Frow has proposed the term *universe of discourse* as an alternative to discursive formation. He gives as examples of universes of discourse 'the religious, scientific, pragmatic, technical everyday, literary, legal, philosophical, magical, and so on', and distinguishes these from *genres of discourse*, which, after Vološinov, he defines as 'normatively structured clusters of formal, contextual, and thematic features, "ways of speaking" in a particular situation' [1986, 67].)

According to Foucault

> Whenever one can describe, between a number of statements, such a system of dispersion, whenever, between objects, types of statement, concepts, or thematic choices, one can define a regularity (and order, correlations, positions and functionings, transformations), we will say, for the sake of convenience, that we are dealing with a *discursive formation* . . . (1972, 38)

All societies, following Foucault, have procedures whereby the production of discourses is controlled, selected, organized and redistributed, and the purpose of these processes of discourse control is to ward off 'powers and dangers' (1981, 52). These procedures govern, variously, what Foucault terms *discursive practices*, *discursive objects*, and *discursive strategies*, such that in all discourses *discursive regularities* can be observed. As Paul A. Bové puts it in his discussion of Foucault's use of the term, discourse 'makes possible disciplines and institutions which, in turn, sustain and distribute those discourses' (Lentricchia & McLaughlin 1990, 57). However, Lynda Nead argues that Foucault is not consistent in his use of the term 'discourse', and that consequently there is some uncertainty about its precise meaning even as it is used in a single work of his (she cites *The History of Sexuality*) (1988, 4).

The work of Mikhail Bakhtin gives us yet further examples of the pressing of the word *discourse* into new services. According to the glossary provided in Bakhtin (1981), *discourse* is used to translate the Russian word *slovo*, which can mean either an individual word, or a method of using words that presumes a type of authority (1981, 427). This is quite close to the usage argued for by Foucault, and this similarity can also be seen in some cognate terms used by Bakhtin. Thus *authoritative discourse* is the privileged language that 'approaches us from without; it is distanced, taboo, and permits no play with its framing context' (1981, 424). In contrast, *internally persuasive discourse* is discourse which uses one's own words, which does not present itself as 'other', as the representative of an alien power. *Ennobled discourse* is discourse which has been made more 'literary' and elevated, less accessible. Tzvetan Todorov gives a number of brief quotations from Bakhtin which show, however, that even his use of the term (or its Russian near-equivalent) has its variations: 'Discourse, that is language in its concrete and living totality'; '*discourse*, that is language as a concrete total phenomenon'; '*discourse*, that is utterance (*vyskazyvanie*)' (Todorov 1984, 25). In his *Problems of Dostoevsky's Poetics*, Bakhtin refers to '*discourse*, that is, language in its concrete living totality, and not language as the specific object of linguistics, something arrived at through a completely legitimate and necessary abstraction from various aspects of the concrete life of the word' (Bakhtin 1984, 181). In the same work, Bakhtin also refers to 'double-voiced discourse', which he claims always arises under conditions of dialogic interaction (1984, 185). It seems clear that IDEOLOGY, variously defined, is a near neighbour to discourse in both Foucault's and Bakhtin's understanding of the term.

For *monovalent discourse* and *polyvalent discourse* (Todorov), see the entry for REGISTER.

See also ARCHAEOLOGY OF KNOWLEDGE; ENUNCIATION; FREE INDIRECT DISCOURSE; NEW HISTORICISM AND CULTURAL MATERIALISM; and the discussion of the difference between text and discourse in the entry for TEXT AND WORK.

Discursive practice See DISCOURSE

Displacement Within Linguistics, displacement refers to the human ability to refer to things removed from the utterer's immediate situation, either in time or in space. This ability seems to distinguish human beings from other living creatures, and it is language-dependent: it is human language which allows us this unparalleled freedom to refer beyond our geographical, social, CULTURAL or historical here-and-now. Thus a statement such as 'Do you remember the nice time we had on holiday in Bulgaria last year?' may appear trivial, but it exemplifies a resource not available in any significant way to other species.

In one sense literature is one of the most sophisticated exemplifications of this ability: members of other species can sham and mislead, but fiction seems a specifically human resource. This should perhaps lead us to consider whether or not literature and other imaginative arts actually play a more important rôle in human development than has sometimes been accorded them. It may also suggest that REDUCTIONIST views of literature as merely a form of reflection of the life-situation of the writer or his or her society are inadequate: clearly what we imagine is based upon what we know, but in important respects it is not limited to it. Our ability to imagine 'what may be the case' or 'what might have been the case' have arguably increased our chances of survival (as well as having given us a much greater ability to endanger ourselves and our future).

Sigmund Freud's use of this term is rather different: see CONDENSATION AND DISPLACEMENT.

Displayed See DEFAMILIARIZATION

Dispositif A term used by Michel Foucault to describe the totality of 'the said and the unsaid' in its heterogeneity. Foucault includes under this heading DISCOURSES, institutions, architectural forms, regulatory decisions, laws, administrative measures, scientific statements, and philosophical, moral and philanthropic propositions (1980a, 194). Sometimes rendered into English as *apparatus*. (See APPARATUS for the Brechtian meaning of this term.)

Disruption See DEFORMATION

Dissemination From Jacques Derrida's book *Dissemination* (1981a), the term describes that state of endless seeding and potential growth of MEANING said to characterize the play of SIGNIFIERS in the absence of SIGNIFIEDS. According to Gayatri Chakravorty Spivak, the translator of Derrida's *Of Grammatology*, the term refers to 'the seed that neither inseminates nor is recovered by the father, but is scattered abroad' (Derrida 1976, xi). Dissemination differs from Empsonian AMBIGUITY to the extent that the flow of new meanings can never be exhausted, nor can these be in any way attached to an AUTHOR: they are the product of language itself.

Dissonant closure See CLOSURE

Dissonant psycho-narration See FREE INDIRECT DISCOURSE

Distance A term used with a range of related meanings in literary criticism, mostly within the theory of NARRATIVE. The most general meaning refers to READER involvement in a literary WORK. Thus whereas readers of Dickens's novels typically became (and become) very involved in the fate of CHARACTERS in the course of reading (gathering on the quay in the United States to meet the latest instalment of *The Old Curiosity Shop* to learn of the fate of Little Nell, for example), much of Joseph Conrad's fiction encourages the reader to observe characters and events more dispassionately, at more of an emotional distance.

This is not unrelated to more specific usages of the term within narrative theory. The reader of Conrad's 'An Outpost of Progress' feels relatively distanced from the fates of the characters of the work, but this is in part because the NARRATOR also seems detached and, if pitying, at a distance from the characters. Distance can, therefore, refer to the gap between STORY and NARRATION, and this gap can be temporal, geographical, or emotional – or traceable to a clash between the value-systems associated with characters and with the narrative. It can also be attributed to more technical matters: the more anonymous and covert the narration, the less distance there is between story and narration; the more the narrative draws attention to itself as narrative (through, for example, a personified narrator), then the greater the story-narrative distance. It should be remembered, however, that it is possible for the technical distance between narrative and story to be very small without necessarily producing very much reader or NARRATEE involvement in the work on an emotional level.

Mieke Bal uses distance as a measure of types of ANACHRONY: the more an event presented in an anachrony is separated from the 'present' (that is, the moment at which the story is interrupted by the anachrony), the greater the distance (1985, 59).

See also ANALEPSIS.

Dominant According to the PRAGUE SCHOOL theorist Jan Mukařovský:

> The systematic foregrounding of components in a work of poetry consists in the gradation of the interrelationships of these components, that is, in their mutual subordination and superordination. The component highest in the hierarchy becomes the dominant. All other components, foregrounded or not, are evaluated from the standpoint of the dominant. (1964, 20)

And Roman Jakobson proposes that

> The dominant may be defined as the focusing component of a work of art: it rules, determines, and transforms the remaining components. It is the dominant which guarantees the integrity of the structure. (1971, 82)

Jakobson goes on to suggest that one may seek a dominant not only in an individual artist's poetic WORK or in the poetic CANON, 'but also in the art of a given epoch, viewed as a particular whole (1971, 83), and he suggests that in Renaissance art 'such an acme of the aesthetic criteria of the time, was represented by the visual arts' (1971, 83).

Brian McHale has pointed out that it is thanks to Roman Jakobson that the concept is known today, but that Jurij Tynjanov is probably the person who deserves to be credited with its invention (1987, 6).

Mikhail Bakhtin makes very extensive use of the concept, especially in his study of Dostoevsky (Bakhtin 1984).

See also CULTURE (dominant culture); DEFAMILIARIZATION; DEFORMATION; HEGEMONY; IDEOLOGY (dominant ideology).

Dominant discourse See DISCOURSE; IDEOLOGY

Dominant ideology See IDEOLOGY

Double-bind A term from interpersonal psychology. In their book *Pragmatics of Human Communication* Paul Watzlawick *et al.* give the following definition.

A double-bind is a message so structured that it both (i) asserts something, and (ii) asserts something about its own assertion such that (iii) assertion *i* contradicts assertion *ii*. Furthermore, it is necessary that the recipient of the message be incapable of stepping outside the FRAME of contradiction set up by these conflicting assertions, such that he or she oscillates between them but cannot resolve the contradiction by means of some sort of meta-assertion (1968, 212).

In his *The Anxiety of Influence*, Harold Bloom invokes the double-bind concept to characterize 'the paradox of the precursor's implicit charge to the ephebe'. For Bloom, what the precursor's poem says to its descendant poem (here Bloom shifts the ground of his argument a little), is, 'Be like me but unlike me' (1973, 70). The EPHEBE'S relation to his precursor, then, is for Bloom essentially neurotic and pathological in nature.

For a more detailed account of Bloom's position see the entry for REVISIONISM.

Duration In NARRATIVE theory, duration can refer either to the time covered by the STORY or part of it (an EVENT), or to the 'time' allotted to either by the TEXT (story-time and text-time). As Rimmon-Kenan points out, the latter concept is a highly problematic one as 'there is no way of measuring text-duration' (1983, 51). On a very rough basis we may note that three years of story-time may be covered by three pages of text, while further on in the same text one hour of story-time may occupy fifty pages. But 'pages' do not give a particularly reliable measure: not only do some READERS read more quickly than others, the same reader will read more or less quickly depending upon such factors as textual complexity, reader involvement and tension, and so on.

Gerald Prince notes that as a result many writers on narrative find *speed* or *tempo* more fruitful concepts in the analysis of narrative texts (1988, 24).

According to Gérard Genette, the four basic forms of narrative movement are ELLIPSIS, pause, scene and summary. These constitute four different ways of varying duration (1980, 94).

See also ISOCHRONY.

Durative anachrony See ANACHRONY

E

Echolalia A term intended to convey the ceaseless echoing back and forth between SIGNS whose significance is determined only relationally and not by any over-riding PRESENCE or fixed authority.

Ecological validity A claim or procedure which lacks ecological validity is one that is based on too narrow or artificial a set of assumptions or test procedures. The term is used to indicate that laboratory experiments may not accurately duplicate or replicate those conditions in 'real life' for which the experiment in question is designed.

Thus for example early behaviourist experiments into the 'effects' of TV programmes which asked schoolchildren a set of questions to determine attitudinal orientation, and then repeated the same questions after the viewing of a TV programme, can be said to lack ecological validity. The way TV alters attitudes is far more subtle and mediated than is presupposed by such experiments.

Within literary criticism and theory the term is most used in the study of READERS AND READINGS. Attempts to determine how 'Common Readers' read are plagued with problems relating to artificial test situations. Ordinary reading is relatively unself-conscious and it intermeshes with a range of other quotidian activities; as soon as an individual is asked questions concerning his or her reading these conditions may be altered in a way that makes the results of such an investigation unreliable.

Écriture French-English dictionaries give 'writing' as the equivalent of écriture, and in an article on DECONSTRUCTIVE criticism M. H. Abrams has glossed écriture as 'the written or printed text' (1977, 428), but the fact that many critics writing in English continue to use the French term suggests that this equivalence is very incomplete. We need to understand that the contemporary critical use of this term dates from Roland Barthes's extension of the meaning of the French term in his *Le Degré Zéro de L'Écriture*, which was published in 1953. The English translation (Barthes 1967b) has what, given the following comments, is arguably the misleadingly title *Writing Degree Zero*. In an article on écriture, Ann Banfield has named other 'landmark texts' which have contributed to the establishment of this term: Maurice Blanchot's 'The Narrative Voice' (1981); Michel Butor, 'L'Usage des Pronoms Personnels dans le Roman' (1964), and Michel Foucault, 'What is an Author?' (1980b) (Banfield 1985, 2). Barthes's translators point out in a note that although in everyday French écriture normally means only 'handwriting', or 'the art of writing', 'It is used here in a strictly technical sense to denote a new concept' (Barthes 1967b, 7). This 'new concept' has to be explained by reference to Barthes's setting of écriture in opposition to *littérature*, a distinction related to that which he makes between *lisible* and *scriptible* or, as rendered in English translation, READERLY AND WRITERLY TEXTS. As Banfield points out, the distinction between écriture and littérature is the more striking in French because prose fiction in French has appropriated to itself certain *grammatical* characteristics which distinguish it from other forms of writing, notably the *passé simple* and the third person NARRATIVE (Banfield 1985, 4). If *littérature* is characterized by these overt grammatical markers, and by less overt and related IDEOLOGICAL ones, *écriture* seeks to escape from 'literariness' by a 'zero style' first, and most strikingly, seen in the French novel in Albert Camus's *L'Étranger*, a novel told in the first and not the third person, and using not the *passé simple* but the *parfait composé*, a grammatical choice which has (or had) a shock effect in French but which is lost in English translation. For Barthes, 'writing degree zero' is a 'colourless writing, freed from all bondage to a pre-ordained state of language' (1967b, 82), it represents an attempt 'to go beyond Literature by entrusting one's fate to a sort of basic speech, equally far from living languages and from literary language proper' (1967b, 83).

What, then, is the significance of this for non-French readers? Banfield argues that although the grammatical markers of écriture are not so apparent in English, nonetheless the term points to a writing characterized by ABSENCE, an absence of the marks of literature, of human agency, which is not limited to French language or CULTURE. Écriture as substantive, says Banfield, 'is a product now divorced from the person and activity of its producer', it is 'the name for the coming to language of a knowledge which is not personal' (Banfield 1985, 13), and she links it with the use of *style indirecte libre*, or FREE INDIRECT DISCOURSE in the novel.

The same loss of complexities can accompany the translating of Jacques Derrida's use of the term *écriture* by the English term *writing*. In an interview with Henri Ronse, published in *Positions*, Ronse puts a case for this complexity to Derrida, who does not reject Ronse's argument.

> *Ronse*: In your essays at least two meanings of the word 'writing' are discernible: the accepted meaning, which opposes (phonetic) writing to the speech that it allegedly represents (but you show that there is no purely phonetic writing), and a more radical meaning that determines writing in general, before any tie to what glossematics calls an 'expressive substance'; this more radical meaning would be the common root of writing and speech. The treatment accorded to writing in the accepted sense serves as a revelatory index of the repression to which archi-writing is subject. (Derrida 1981b, 7–8)

For *archi-writing*, see the entry for ARCHE-WRITING. It seems clear that for Derrida, *écriture* and *arche-writing* can on occasions perform almost interchangeable rôles.

Theories of écriture clearly appealed to STRUCTURALISTS who were keen to stress the depersonalized systematics of LANGUE and eager to reject views which saw language always in terms of (normally individual) human origins or PRESENCE. So far as literature is concerned, an adoption of the term seems typically to stem from impulses similar to those which lie behind the belief in the death of the AUTHOR, and its use often betokens a commitment to a writerless writing, or at least to a writing seen as intransitive and transparent.

See also ÉCRITURE FÉMININE.

Écriture féminine According to Elaine Showalter, 'the inscription of the feminine BODY and female difference in language and text' (1986, 249). Showalter's discussion of the concept is worth consulting in its entirety. The term was coined by French FEMINISTS, and represents more a description of an ideal, future achievement than of a particular type of writing of which there already exist many examples.

The name most frequently associated with the term is that of Hélène Cixous, although those described as practitioners of écriture féminine have, themselves, rarely used the term. Nor can one expect a tidy definition or theorization of the term from Cixous; as she writes in one of her best-known pieces, 'The Laugh of the Medusa':

> It is impossible to *define* a feminine practice of writing, and this is an impossibility that will remain, for this practice can never be theorized, enclosed, coded – which doesn't mean that it doesn't exist. But it will always surpass the discourse

> that regulates the phallocentric system; it does and will take place in areas other than those subordinated to philosophico-theoretical domination. It will be conceived of only by subjects who are breakers of automatisms, by peripheral figures that no authority can ever subjugate. (1981, 253)

The concept has interesting forebears. Take the following comment from Virginia Woolf's essay 'Women and Fiction', first published in 1929.

> But it is still true that before a woman can write exactly as she wishes to write, she has many difficulties to face. To begin with, there is the technical difficulty – so simple, apparently; in reality, so baffling – that the very form of the sentence does not fit her. It is a sentence made by men; it is too loose, too heavy, too pompous for a woman's use. (1966b, 145)

Woolf concludes that a woman must 'alter and adapt' the current sentence until she can write one that 'takes the natural shape of her thought without crushing or distorting it' (1966b, 145).

In another essay she makes it clear that it is not just the shape of a woman's thought but the reality of her body that manifests itself as a problem of writing for her. In 'Professions for Women', which was first read as a paper to the Women's Service League, she pictures for her listeners a girl sitting with pen in hand, and notes that the image that this picture brings to mind is that of a fisherman 'lying sunk in dreams on the verge of a deep lake with a rod held out over the water'.

> Now came the experience that I believe to be far commoner with women writers than with men. The line raced through the girl's fingers. Her imagination had rushed away. It had sought the pools, the depths, the dark places where the largest fish slumber. And then there was a smash. There was an explosion. There was foam and confusion. The imagination had dashed itself against something hard. The girl was roused from her dream. She was indeed in a state of the most acute and difficult distress. To speak without figure, she had thought of something, something about the body, about the passions which it was unfitting for her as a woman to say. (1966a, 287–8)

More recent feminist accounts of *écriture féminine* suggest that this sense of one's own body may become the source from which the new writing must stem. Madeleine Gagnon, for example, after noting that she has to take over a language which, although it is hers, is foreign to her, argues that there is an alternative.

> All we have to do is let the body flow, from the inside; all we have to do is erase, as we did on the slate, whatever may hinder or harm the new forms of writing; we retain whatever fits, whatever suits us. (1980, 180)

Some feminist commentators have, however, suggested that the process may not be so easy as is here assumed, pointing out that women's sense of their own bodies may already be saturated with IDEOLOGICALLY foreign elements: what flows from the inside of the body may thus be more than body alone.

Ellipsis Alternatively *gap*. The omitting of one or more items in a NARRATIVE series: any gap of information in a temporal or other sequence. We never learn anything concrete, for example, of Heathcliff's history prior to his discovery in Liverpool by Mr Earnshaw, nor of what he does to become rich between the time of his disappearance and re-appearance in *Wuthering Heights*. In this case the ellipsis is relatively *unmarked* (or implicit), as it covers information not known to the personified NARRATORS. But when in Charles Dickens's *Bleak House* Esther Summerson seeks to explain why she finds Mrs Woodcourt irksome, and breaks off with the words, 'I don't know what it was. Or at least if I do, now, I thought I did not then. Or at least – but it don't matter', then we have a clearly *marked* (or explicit) ellipsis: the READER's attention is drawn to the fact that something that is known to the narrator is withheld from him or her. (See the entry for MARKEDNESS.)

Gérard Genette characterizes certain ellipses as *hypothetical*; these are those ellipses which are impossible to localize or – on occasions – to place in any spot at all, but which are revealed after the event by an ANALEPSIS (1980, 109).

Ellipses can be permanent or temporary: in most detective novels certain marked gaps are sustained until the end of the WORK only to be filled in the course of the final pages.

In a related but different sense, theorists such as Roman Ingarden and Wolfgang Iser have drawn attention to the gaps and indeterminacies necessarily possessed by literary works, gaps and indeterminacies which must be filled in or fleshed out and, as it were, CONCRETIZED by readers.

See also ABSENCE; narrative movements (entry for DURATION); PARALIPSIS.

Embedded event See EVENT

Embedded narrative See FRAME

Embedding Alternatively nesting or staircasing. In general terms, embedding involves the enclosing of one unit in a larger unit – a clause in a sentence, a sub-clause in a clause, and so on. Within different branches of Linguistics the terms *embedding* and *nesting* may be given varying meanings, often relating to the extent to which the enclosed unit either determines or modifies the enclosing one.

Similarly, in NARRATIVE embedding refers in its simplest form to what is often called 'a story within a story': Dickens's *The Pickwick Papers* contains a succession of embedded tales told by Sam Weller. But certain narrative theorists use the term only when, as above, the enclosed unit (which may not be a complete story) either determines or modifies the enclosing unit in some way. Thus Shlomith Rimmon-Kenan uses embedding to translate Claude Bremond's term *enclave*, and she points out that he uses this to indicate where one narrative sequence is 'inserted into another as a specification or detailing of one of its functions' (Rimmon-Kenan 1983, 23).

See also EVENT (embedded event) and FRAME (embedded frame).

Emotive See FUNCTIONS OF LANGUAGE

Empirical / empiricism See SOLUTION FROM ABOVE / BELOW

Empirical reader See READERS AND READING

Empiricist fallacy See NORM

Emplotment A term much used within NEW HISTORICIST writing which refers to the textualizing of historical 'facts' to create a plot or NARRATIVE. Just as a novelist may create any amount of novels from a given set of CHARACTERS and events, so too (certain radical historians have argued) a given set of historical data can be emplotted in innumerable manners, all of which will present that material in different ways.

Use of the term often draws on the distinction between STORY and PLOT to be found in narrative theory, suggesting that just as we can get at a story only through a plot, so too it is an illusion to believe that the raw facts of history can be directly apprehended, as they too must be emplotted in one form or another. The implication is the process of emplotment should be made opaque rather than transparent, that writers of history should come clean about the principles around which they are structuring their narratives of history.

In their most extreme form such arguments espouse highly idealistic positions which deny the existence of historical facts or events as such, and argue that all the historian has is a set of different emplotments which cannot be judged against a reality that each claims to represent or reproduce. Here Jacques Derrida's notorious claim that there is nothing outside the TEXT is taken to a logical extreme.

Enchained event See EVENT

Energy Otherwise *social energy*. A term associated with the NEW HISTORICISTS and especially with Stephen J. Greenblatt, who provides an account of the origin and significance of this term in 'The Circulation of Social Energy' (Greenblatt 1988, 1–20).

According to Greenblatt, he wished to find out why in Renaissance England cultural objects, expressions and practices – and, especially, plays by Shakespeare – acquired compelling force. Turning to English literary theorists of that period he found that they needed a new word for this force,

> a word to describe the ability of language, in Puttenham's phrase, to cause 'a stir to the mind'; drawing on the Greek rhetorical tradition, they called it *energia*. This is the origin in our language of the term 'energy,' a term I propose we use, provided we understand that its origins lie in rhetoric rather than physics and that its significance is social and historical. We experience that energy within ourselves, but its contemporary existence depends upon an irregular chain of historical transactions that leads back to the late sixteenth and early seventeenth centuries. (1988, 4–5)

This does not mean that when modern spectators feel the aesthetic force of an Elizabethan play they are experiencing what a contemporary of the playwright's would have experienced. Continuous processes of NEGOTIATION and EXCHANGE have reformulated this aesthetic force anew for each succeeding generation; the play's social energy is still active, but transformed in its effects from what these were several centuries ago. Indeed, one of the things which characterizes art is that its social energy can survive transplantation to new social and historical contexts.

Énoncé / Énonciation See ENUNCIATION

Enthymeme According to V. N. Vološinov, 'every utterance in the business of life is an objective social enthymeme. It is something like a password known only to those who belong to the same social purview' (1976, 101).

John Frow has developed this argument using Althusser's concept of INTERPELLATION. According to Frow the concept of presupposition is central to a theory of how implied subject positions are locked into implied structures of meaning, and Vološinov's conception of 'the *enthymematic* structure of discourse defines the logic of self evidence which is an important consequence of generic norms'. For Frow, the 'free' or implicit information in a statement is often more important than its 'tied' or overt information, as this 'free' information 'anchors the statement to a context other than the immediate one' (1986, 77–8). A genre thus may carry with it important IDEOLOGICAL constraints and determinations.

Entrapment See INCORPORATION

Enunciatee See ENUNCIATION

Enunciation Along with cognate words such as *enunciatee*, *enunciator* and *enunciated*, enunciation has now begun to replace the French loan-words *énonciation*, *énonciateur*, and *énoncé*, but such translations often fail to carry the more specific meanings of the French originals. (The translator of Roland Barthes's article 'To Write: an Intransitive Verb?', for example, regularly includes the French original terms alongside the English translation, and *énonciation* is translated on one occasion as *utterance* and on another as *statement*.

Umberto Eco, in contrast, renders énoncé and énonciation as *sentence* and *utterance* [1981, 16].)

What is central to use of the various French terms is a distinction between the particular, time-bound *act* of making a statement, and the *verbal result* of that act, a result which escapes from the moment of time and from the possession of the person responsible for the act. We can note that the important distinction between *utterance* and *statement* is that the former term links that uttered to its human originator, whereas the latter term concentrates attention on to the verbal entity itself. When *énonciation* is used in French it more usually has the meaning we attribute to *utterance*, that is to say, it calls to mind the *act* of producing a form of words which involves a human SUBJECT. In contrast, when *énoncé* is used the intention is normally to consider a form of words independently from their association with a human subject.

In addition, the French terms generally include the idea of a human target or audience (whereas a statement or UTTERANCE can be made in the absence of these). Thus some writers in English prefer to translate the term *énonciateur* as *addresser*, a usage which inevitably perhaps also brings with it *addressee* – the person at whom an utterance is aimed. When one meets with the term *enunciation* in English care should therefore be taken. This often implies the precise act-in-a-context which produces a person-oriented utterance, but it can also have a more restricted sense of the *evidence* remaining in an utterance that it stems from a *subject's act-in-a-context.*

See also the DISCOURSE / STORY distinction in the entry for discourse.

Ephebe According to the OED, in Classical Greece a young citizen aged between 18 and 20 years who was chiefly involved in garrison duty. The term has been pressed into new service by Harold Bloom to describe the 'young citizen of poetry' or 'figure of the youth as virile poet' (1973, 10, 31). Bloom's ephebe, it should be said, seems to be engaged more in storming the paternal garrison than in defending it.

Episodic narrative See STRING OF PEARLS NARRATIVE

Épistémè (Sometimes written *epistēmē*.) A term coined by Michel Foucault (based on the Greek for knowledge) and widely used by, among others, Jacques Derrida, to indicate the totality of relations and laws of transformation uniting all discursive practices (see the entry for DISCOURSE) at any moment of time.

The term can, alternatively, refer to a given historical period during which the above-mentioned relations and laws of transformation are constant and stable. Thus *épistémè* has points of contact with Marx's *ruling ideas* and with the MARXIST sense of IDEOLOGY, but it has a more all-embracing, totalizing sense: an *épistémè* leaves no room – or attempts to exclude the space – for any ways of producing or arranging knowledge apart from its own. As Richard Harland points out, one problem caused for Foucault by the theory of the *épistémè* is that the theorist of the concept (i.e. Foucault himself) must be part of the *épistémè* if indeed the *épistémè* is all-embracing (Harland 1987, 123). A further set of problems involves the reason why – and manner whereby – one *épistémè* gives way to and is replaced by another.

See also the entries for PARADIGM SHIFT and PROBLEMATIC, for comparable concepts.

Epistemological break Epistemology is the theoretical study of knowledge – its nature, how it is to be studied or achieved, what its grounds are. The concept of the *epistemological break* is associated with the theories of the French MARXIST philosopher Louis Althusser, who developed it initially to describe what he saw as the significance of the changes reflected by 'the confrontation between Marx's Early Works and *Capital*' (1969, 13). For Althusser, this confrontation involved a major opposition, the opposition that separates science from IDEOLOGY. Althusser's definition of these two terms is not uncontroversial, and has been subjected to considerable criticism in recent years. It involves, among other things, attributing to science a self-validating rôle which, as Terry Eagleton has pointed out, runs counter to most traditions of Marxist thought (1991, 137–8).

See also COPERNICAN REVOLUTION; *ÉPISTÉMÈ*; PARADIGM SHIFT; PROBLEMATIC.

Epoché In PHENOMENOLOGICAL criticism, that suspension of all pre-existing beliefs and attitudes which must precede the analysis of consciousness. According to Gérard Genette,

> when the semiologist has operated the semiological reduction, the *epoché* of meaning on the object-form, he is presented with a matte object, cleansed of all the varnish of dubious, abusive significations, with which social speech had covered it, restored to its essential freshness and solitude. (1982, 39)

Erasure In recent theoretical usage, normally associated with a practice popularized by Jacques Derrida and the bane of typesetters and proof readers, of leaving deleted words 'under erasure' (*sous rature*) in his writings – that is, of leaving them crossed out but not removed. Derrida apparently adopted the practice after noting Martin Heidegger's use of it. The erasure marks thus act in a manner similar to what are known colloquially as scare quotes; that is, quotation marks placed round a word with the similar intention of drawing attention to its inadequacy or questionable validity.

See also the entry for ARCHE-WRITING.

The term is sometimes given a more general meaning in discussion of MODERNIST and POSTMODERNIST techniques whereby achieved verisimilitude or REALISM is subsequently denied. Thus an erased CHARACTER would be a character who the reader accepted as 'real' according to realist CONVENTIONS, but who was shown later to be not-real in some sense, either within the achieved world of the fiction or by a transference of the character from this world to the extra-fictional world of the text. (See the discussion in McHale 1987, 64–6, and compare CANCELLED CHARACTER).

Certain FEMINIST writers have extended the term to describe the way in which women are rendered invisible by PATRIARCHAL accounts which effectively treat them as invisible or non-existent. In his introduction to a recent special issue of the journal *The Conradian* concerned with 'Conrad and Gender', for example, the editor, Andrew Michael Roberts, refers to a reference-book summary of the plot of *Lord Jim* which omits all mention of the character Jewel – a character already MUTED in her relations with other characters in the novel (Roberts 1993, vi).

Erlebte Rede See FREE INDIRECT DISCOURSE

Essentialism The belief that qualities are inherent in objects of study, and that therefore the contexts in which these exist or are studied are irrelevant. Essentialism is therefore to be distinguished from DIALECTICAL, contextual or relational theories and approaches. The term often carries with it the implication that the qualities of objects of study are self-evident and do not themselves need to be sought for or explained (see Cameron 1985, 187).

Defenders of HUMANISM are often accused of essentialism by their critics, and Alan Sinfield argues that

> The essentialist-humanist approach to literature and sexual politics depends upon the belief that the individual is the probable, indeed necessary, source of truth and meaning. Literary significance and personal significance seem to derive from and speak to individual consciousnesses. But thinking of ourselves as essentially individual tends to efface processes of cultural production and, in the same movement, leads us to imagine ourselves to be autonomous, self-determining. (1992, 37)

The opposite of essentialism is RELATIVISM, but it is usual for both of these terms to be used pejoratively. The term 'relationism' is sometimes used as a non-pejorative alternative to relativism.

Event According to Mieke Bal an event is 'the transition from one state to another state' in a NARRATIVE (1985, 5).

Steven Cohan and Linda M. Shires distinguish between *kernel* and *satellite* events: the former 'advance or outline a sequence of transformations' while the latter 'amplify or fill in the outline of a sequence by maintaining, retarding, or prolonging the kernel events they accompany or surround' (1988, 54). They further argue that events are *enchained* when they are found in back-to-back succession with each other, whereas it is, alternatively, possible for one event to be *embedded* in another (1988, 57; such EMBEDDING is also called *stair-casing*). Finally, they point out that an event can be *singular*, *repeated*, or *iterative* (1988, 86).

Roland Barthes (1975) makes a similar distinction between *catalyses* and *nuclei*: the former denotes events that are not logically essential to the narrative, and the latter those that are.

Compare *kernel function* in the entry for FUNCTION, and see also the entry for FREQUENCY.

Exchange A term used by the NEW HISTORICIST writer Stephen J. Greenblatt to describe the manner whereby works of art *negotiate* new forms of SOCIAL ENERGY as they enter the life of social and cultural contexts different from those in which they were created. Talking about such negotiations in his book *Shakespearian Negotiations* (1990; first published 1988), Greenblatt describes his search for 'an originary moment' in which 'the master hand shapes the concentrated social energy into the sublime aesthetic object'. The search yields no such moment, however.

> In place of a blazing genesis, one begins to glimpse something that seems at first far less spectacular: a subtle, elusive set of exchanges, a network of trades and trade-offs, a jostling of competing representations, a negotiation between joint-stock companies. (1990, 7)

Commentators have drawn attention to the fact that Greenblatt's metaphor-kitty seems to be supplied almost exclusively by the stock-market here, and some have argued that with these metaphors Greenblatt imports a set of values and assumptions that lead to one sort of EMPLOTMENT rather than another.

Exchange value See FETISHISM

Exotopy The word suggested in the English translation of Tzvetan Todorov's *Mikhail Bakhtin: The Dialogical Principle* (1984) to represent a coinage of Bakhtin's which describes an AUTHOR's movement outside of and away from his or her CHARACTER subsequent to an earlier, initial stage of identification and empathizing with the character.

Both movements are important to Bakhtin: the novelist must understand a character from within, but in order to understand the character fully it must be perceived as OTHER, as apart from its creator and in its distinct *alterity*. Moreover, DIALOGUE is only possible with an 'other': one can only talk to oneself if one estranges oneself from part of oneself and treats this part as other.

Compare CANCELLED CHARACTER.

Expressive(s) See FUNCTIONS OF LANGUAGE; SPEECH ACT THEORY

Extent See ANALEPSIS

Exteriority In Michel Foucault's usage, rejecting the procedure whereby the investigator proceeds from DISCOURSE to its 'interior, hidden nucleus, towards the heart of a thought or a signification supposed to be manifested in it', and instead, adopting the alternative procedure whereby the investigator proceeds on the basis of discourse itself, 'its appearance and its regularity, [and moves] towards its external conditions of possibility, towards what gives rise to the aleatory series of these events, and fixes its limits' (1981, 67).

This recommendation ties in with a number of movements in recent theory which reject the search for a hidden, inner, CENTRE or PRESENCE and instead seek to explain the object of study in terms of the possibilities engendered by its existence in a complex of shifting relations. In literary criticism an analogous movement would be represented by the rejection of ESSENTIALIST views of the literary TEXT in favour of a study of the literary text's 'conditions of possibility' in different READING or INTERPRET[AT]IVE COMMUNITIES.

See also RADICAL ALTERITY for Jacques Derrida's use of this term.

Extradiegetic See DIEGESIS AND MIMESIS

Fabula See STORY AND PLOT

Fabulation See MODERNISM AND POSTMODERNISM

Face-threatening acts (FTAs) See POLITENESS

Fantastic An early concern with the fantastic can be found in the work of the RUSSIAN FORMALISTS. In his essay 'Thematics', for example, Boris Tomashevsky quotes an interesting passage from Vladimir Solovyev's introduction to Alexey Tolstoy's novel *The Vampire*, which Tomashevsky describes as 'an unusually clear example of fantasy'. According to Solovyev, the distinguishing characteristic of the genuinely fantastic is that

> it is never, so to speak, in full view. Its presence must never compel belief in a mystic interpretation of a vital event; it must rather point, or *hint*, at it. In the really fantastic, the external, formal possibility of a simple explanation of ordinary and commonplace connections among the phenomena always remains. This external explanation, however, finally loses its internal probability. (Tomashevsky 1965, 83–4)

Does the fantastic constitute an independent genre, and, if so, what its defining characteristics? Perhaps the most influential contribution to the view that the fantastic constitutes such an independent genre is Tzvetan Todorov's *The Fantastic: A Structural Approach to a Literary Genre* (1973). Christine

Brooke-Rose has made a useful summary of the three conditions which Todorov believes to be more or less standard components of the 'pure' fantastic. The READER must hesitate between natural and supernatural explanations of what happens in the WORK up to its conclusion; this hesitation may be represented – that is, it may be shared by a leading CHARACTER in the work (this, according to Todorov, is normal but not essential); and the reader must reject both a poetic and an allegorical reading of the work, as both of these destroy the hesitation which is fundamental to the pure fantastic (Brooke-Rose 1981, 63). If there is no hesitation, then either we are in the realm of some variant of the *uncanny* (the events are seen by the reader to have a natural explanation), or of the *marvellous* (the events are seen by the reader to have a supernatural explanation). There would seem here to be full agreement with Solovyev's above-quoted definition. (For the uncanny, see Sigmund Freud's essay 'The Uncanny' [Freud 1955], in which Freud relates the uncanny [or *unheimlich*] to 'that class of the frightening which leads back to what is known of old and long familiar' [1955, 220], but which has become alienated from the mind through the process of repression [1955, 241]. Through an analysis of E. T. A. Hoffmann's tale 'The Sandman', Freud reaches the conclusion that a major component of the uncanny is the fear of castration.)

Brooke-Rose points out that Todorov's rather demanding conditions leave us with very few examples of the pure fantastic – that is, works in which a hesitation between natural and supernatural explanations lasts until the very end of the STORY. She suggests that the pure fantastic is 'not so much an evanescent *genre* as an evanescent *element*' (1981, 63), which of course makes the fantastic a much more widespread literary phenomenon.

Kathryn Hume, in her *Fantasy and Mimesis* (1984), provides an extended discussion of the problems of defining the fantastic, and in her summary of attempted definitions she categorizes these as either one-, two-, three-, four-, or five-element definitions. The elements involved range from the choice of subject matter, the changing of 'ground-rules' as with Alice's discovery of the new rules governing Wonderland (Erik Rabkin), the 'persuasive establishment and development of an impossibility' (W. R. Irwin), satisfying readers' desire for recovery, escape, consolation (J. R. R. Tolkien), and tracing 'the unsaid and the unseen of culture: that which has been silenced, made invisible' in a way that is fundamentally subversive (Rosemary Jackson) (Hume 1984, 13-17).

Some commentators make an overt or implied distinction between *fantasy* and *the fantastic*. Thus, for example, Anne Cranny-Francis uses *fantasy* as an umbrella-term containing three different sub-types: 'other-world fantasy', 'fairy-tale', and 'horror' (1990, 77). It is clear that in this usage fantasy is different from (if related to) the fantastic.

Faultline By analogy with the physical faultlines around which earthquakes take place because of the pressure between two opposing land plates – any structural weakness or contradiction in an IDEOLOGY or a society from and around which disturbances can be expected.

The term is associated with Alan Sinfield, whose book *Faultlines: Cultural Materialism and the Politics of Dissident Reading* (1992) sets itself the task of exploring a number of such specific faultlines, mainly from the early modern period. Sinfield provides a good example of what he means by the term in a discussion of Shakespeare's *Macbeth.*

> In *Macbeth,* Duncan has the legitimacy but Macbeth is the best fighter. Duncan cannot but delegate power to subordinates, who may turn it back upon him – the initial rebellion is that of the Thane of Cawdor, in whom Duncan says he 'built / An absolute trust.' If the thought of revolt can enter the mind of Cawdor, then it will occur to Macbeth, and others; its source is not just personal (Macbeth's ambition). Of course, it is crucial to the ideology of absolutism to deny that the state suffers such a structural flaw. (1992, 40)

Much of Sinfield's discussion involves the way in which authority tries to conceal such faultlines while dissidence attempts to expose them – often in transposed or disguised forms.

Felicity conditions See SPEECH ACT THEORY

Female affiliation complex A term proposed by Sandra M. Gilbert and Susan Gubar in *The War of the Words*, the first volume of their study *No Man's Land* (1988). They build on Freud's model of the family romance as outlined in his 'Female Sexuality' (1963; first published 1931). In this work Freud suggests that the growing girl may follow one of three lines of development when, as she enters the Oedipal phase, she definitively confronts the fact of her femininity. Either she can turn her back on sexuality altogether, or she can cling in obstinate self-assertion to her threatened masculinity, or, finally, she can arrive at the ultimate normal feminine attitude in which she takes her father as love-object and thus arrives at the Oedipus complex in its feminine form.

Gilbert and Gubar compare the situation of Freud's 'growing girl' to that of the woman writer in the twentieth century, who confronts both a matrilineal and a patrilineal inheritance. Her reaction to these two 'parents' can take three forms similar to those faced by the 'growing girl', but whichever path she chooses, according to Gilbert and Gubar, she will have to struggle with a *female affiliation complex.* Very often the woman writer will oscillate between more than one of the available options, but whatever she does she will experience anxiety about her choice. For this reason, they argue, what is needed is 'a paradigm of ambivalent affiliation, a construct which dramatizes women's intertwined attitudes of anxiety and exuberance about creativity' (1988[i], 170). To this paradigm they give the name of the 'female affiliation complex'.

Feminism Toril Moi makes a useful distinction between three cognate terms which provides a good starting point: *feminism* is a political position, *femaleness* a matter of biology, and *femininity* a set of CULTURALLY defined characteristics (1986, 204). It should be recognized, of course, that Moi's suggested definitions here have a political edge: she is as much arguing for how these terms *should* be used as describing an actual, existing usage. Phrases such as 'the eternal feminine' make it clear that non-feminist usages can define femininity in universal, biological rather than cultural terms – nor is it that uncommon to find 'female' used to refer to culturally acquired characteristics. (The OED definitions of femininity make interesting reading in this context.)

However good a starting point this is it is not unproblematic – not least because Elaine Showalter, in her *A Literature of Their Own*, suggests a different way of using these three terms in the narrower field of women's writing. For Showalter, the feminine stage of women's writing involves a prolonged phase of imitating the prevailing modes of the dominant tradition and internalizing its standards of art; the feminist stage involves the advocacy of minority rights and values; and the female stage is the phase of self-discovery and search for identity (1982, 13).

Of the three terms, feminism is probably the most complex. The OED describes the word as 'rare', and defines it as 'the qualities of females', giving an example from 1851. But from the end of the nineteenth century the word comes increasingly to be applied to those committed to and struggling for equal rights for women – including men: in Joseph Conrad's *Under Western Eyes* (first book publication 1911), for example, the CHARACTER Peter Ivanovitch is repeatedly referred to (ironically) as a feminist. Moreover, not all those women fighting for women's rights accepted the term. In Virginia Woolf's *Three Guineas*, first published in 1938, Woolf writes

> What more fitting than to destroy an old word, a vicious and corrupt word that has done much harm in its day and is now obsolete? The word 'feminist' is the word indicated. That word, according to the dictionary, means 'one who champions the rights of women'. Since the only right, the right to earn a living, has been won, the word no longer has a meaning. (1977, 117)

Woolf argues, too, that the word *feminist* was one which was applied to those fighting 'the tyranny of the patriarchal state', 'to their great resentment' (1977, 118) – in other words, that the word was imposed on rather than chosen by women fighting for the rights of women.

Feminism as socio-political movement experienced a resurgence in the late 1960s and early 1970s, especially in Western Europe and the United States, a resurgence which continues and which has established a number of seemingly permanent changes in the developed countries – and which has not been without an effect in the developing world. Since that time feminism has become more and more of an international movement, with increasing contacts between activists and sympathizers in different parts of the world. From the start of this movement, the rôle of literature was considerable. This is partly because literary writing was less closed to women than most of the other arts, and other forms of writing, but also because the literature of the past written (especially) by women offered itself as a record and analysis of the past oppression of women. It should also be remembered that the modern resurgence of feminism had one of its most important sources in the universities and colleges of the developed world, amongst a group of widely read women.

Radical feminism is a term still current but perhaps more in use in the 1960s and 1970s. It is in its insistence upon the fundamental and all-embracing significance of gender differentiation that radical feminism's radicalness is normally taken to consist – along with (often but not always) a rejection of most or all forms of collaboration with men or with organizations containing men. Radical feminism is often (but again, not always) associated with a commitment to Lesbianism, and if it is possible for a man to be a feminist it seems impossible (or at the very least extremely difficult) for one to be a radical feminist. Radical feminism tends to be universalizing rather than to focus upon the socially, culturally, and historically specific characteristics of PATRIARCHY, although to this it needs to be added that radical feminists have led important campaigns against specific forms of oppression. Eve Kosofsky Sedgwick has commented, in criticism of radical feminism, that it

> tends to deny that the meaning of gender or sexuality has ever significantly changed; and more damagingly, it can make future change appear impossible, or necessarily apocalyptic, even though desirable. Alternatively, it can radically oversimplify the prerequisites for significant change. In addition, history even in the residual, synchronic form of class of racial difference and conflict becomes invisible or excessively coarsened and dichotomized in the universalizing structuralist view. (1993, 13)

Of perhaps most specific interest to students of literature has been the radical feminist analysis of patriarchal and SEXIST elements in language. Representative radical feminists are Adrienne Rich, Mary Daly and Shulamith Firestone.

Fetishism In the MARXIST sense of the term, fetishism is closely linked to Karl Marx's distinction between *use value* and *exchange value*. Marx distinguishes between the value a commodity has measured in terms of what it can be exchanged for (exchange value), and measured in terms of its use to whoever possesses it (use value). Fetishism for Marxists thus involves a confusion between the two: the miser hoards his or her gold as if it had value in itself, not realizing that it is valuable only in terms of what it can be exchanged for. For Marx, *commodity fetishism* is a form of fetishism in which the value given to commodities within a system of exchange – a set of relationships – is incorrectly believed to be intrinsic to the commodity itself. The terms have frequently been brought into discussions of literary value by Marxist critics, especially in the context of discussions of the way in which literature itself becomes a commodity in the eighteenth and nineteenth centuries.

Fetishism plays a rather different rôle in Freudian theory. According to Juliet Mitchell, Freud's theory of fetishism was a long time in the making, for although a number of its crucial aspects are pinpointed quite early in his work, the synthesis comes only in 1927 (1974, 84–5). For Freud, fetishism is connected to the *castration complex* which, according to Freud, follows the boy child's shock at the sight of his mother's genitals. As Mitchell puts it:

> Instead of acknowledging this evidence of castration they set up a fetish which substitutes for the missing phallus of the woman, but in doing this the fetishists have their cake and eat it: they both recognize that women are castrated and deny it, so the fetish is treated with affection and hostility, it represents the *absence* of the phallus and in itself, by its very existence, asserts the *presence* of it.
>
> (1974, 85)

Thus in Freudian theory the fetish is there *to disguise a lack*. Roland Barthes applies this to the literary TEXT: the text itself is a fetish object which disguises the lack of the author (1976, 27). The author is of course, as Barthes has already told us, dead: see AUTHOR.

See also REIFICATION.

Figural See NARRATIVE SITUATION

Figure According to Gérard Genette, there is a *gap* between what the poet has written and what he has thought. Like all gaps, Genette continues, this gap has a form, and the form is called a figure (1982, 47). Genette thus relates the term to some of the traditional concerns of rhetoric; indeed, he rejects the term *figure of thought* on the ground that a figure pertains not to *thought* but to *expression* (1982, 54; compare the more idiomatic English expression *figure of speech*).

See also ABSENCE: FIGURE AND GROUND.

Figure and ground. Experiments have confirmed that the process of visual perception typically involves a sorting of the information received by the brain from the eye into two categories, which psychologists name figure and ground. This process of sorting prevents the brain from being swamped by too much information, and allows for concentration upon certain aspects of incoming messages at the expense of other aspects.

R. L. Gregory has cited the Dane, Edgar Rubin, as the psychologist whose name is associated with the first experiments involved with figure-ground reversal. In 1915 Rubin published a study, *Synoplevende Figurer*, describing his use of line drawings which were ambiguous and could be perceived in two ways: the best-known example of these (much pirated for book jackets!) is that of the picture which can be seen either as an elegant urn or vase, or as two human faces turned towards each other. Gregory points out that such trick pictures tell us something about the active nature of perception and of the principles on which the brain sorts out incoming information (1970, 15–18). What is achieved by Rubin's pictures, or a conjuror's sleight-of-hand, is a disturbing of the perceptual process either in the interests of illustrating some of the principles which govern it, or of trickery.

According to Irvin Rock (1983), the essence of the figure-ground concept

> is not, contrary to popular opinion, that one or another region stands out from the background (although that is true) but that the contour dividing two regions ends up *belonging to* the region that becomes figure. It therefore gives that region rather than the other a specific shape. (1983, 65)

Both Rubin and Rock are concerned with the processes of visual perception, but many recent commentators on the RUSSIAN FORMALIST concepts of FOREGROUNDING and DEFAMILIARIZATION (which have, potentially, a much wider application) have associated both with figure-ground separation; just as we pay only passing attention to what we perceive as 'ground', reserving our attention and concern for what we single out as 'figure', so too there are things – concepts, ideas, attitudes as much as objects – which we perceive as STEREOTYPES in a 'familiarized' manner. And just as Rubin's diagrams upset the fixed categories with which the brain sorts information received from the eye, so literary defamiliarization upsets other fixed categories which allow us to skip over certain information without scrutinizing it at length or in detail. One of the reasons why some literary WORKS give new READING experiences time and time again may well be that we make different figure-ground distinctions on successive readings.

See also FRAME.

Filters See PERSPECTIVE AND VOICE

First-person See NARRATIVE SITUATION

Flashback See ANALEPSIS

Flashforward See PROLEPSIS

Flicker The effect achieved by a certain sort of AMBIGUITY, in which rather than having two clear alternatives between which to choose, the READER is disturbed by flashes of alternative meanings. The term is Brian McHale's, and he relates it to Roman Ingarden's concept of *iridescence* or *opalescence*, found where two alternative worlds are struggling for supremacy in a TEXT but neither is capable of achieving it (1987, 32).

See also FANTASTIC.

Flip-flop According to Bernard Dupriez, a term which has a range of MEANINGS involving symmetrical substitution. This can be at a syntactical level, in which two elements having the same function are exchanged in two syntactically identical sequences: Dupriez illustrates this with a quotation from J. Prévert's *La Pluie et le beau temps*:

> Yes I have a glass leg
> and I have a wooden eye.

But the term can also be used in NARRATOLOGY, and Dupriez gives the example of an O'Henry story in which a young servant girl pretending to be a millionairess meets an extremely rich heir who is pretending to be a waiter (Dupriez 1991, 193).

Focalization See PERSPECTIVE AND VOICE

Folk See POPULAR

Foregrounding See DEFAMILIARIZATION

Foreshadowing See PROLEPSIS

Formulaic literature A concern with the formulaic element in art and literature during the present century is closely connected with investigations into POPULAR and folk art, but has spread beyond the boundaries of such investigations. A key figure here is that of the Russian Vladimir Propp, whose *Morphology of the Russian Folktale* was first published in Russian in 1928. Propp based his work on the study of a corpus of nearly two-hundred Russian folk tales, and attempted to abstract common elements from these, elements which he named FUNCTIONS. Propp's work was ground-breaking, but it focussed attention on to what is rather an obvious point, that folk tales rely very heavily on elements that recur from tale to tale – formulae. (Think how many fairy tales begin with 'Once upon a time' and end with 'And they lived happily ever after'.) We are led to speculate upon two again obvious and related questions: where does this heavy reliance upon formulae come from, and what function does it perform?

It seems that *oral* performance typically involves a heavy reliance upon the formulaic, and that the advent of writing leads to a lessening of the verbal artist's reliance upon formulaic elements. Just as those who must frequently speak in public without notes in our own time tend to rely upon formulaic expressions (think of politicians and the tellers of jokes), so too the oral poet or story-teller needed formulae to act as props to the memory, while his or her listeners also found them useful as they injected familiar elements into that which could not be read a second time and thus made it easier to assimilate. To use a technical term we can say that a reliance upon formulaic elements increases the REDUNDANCY rate, and it is a fact that speech has (and needs) a higher redundancy rate than does writing, which is one reason why popular film and TV is often highly formulaic. This does not mean to say that formulaic elements have no aesthetic significance; as Max Lüthi puts it

> The esthetics of production and the esthetics of reception are parallel, just as mnemetic technique, the basis of oral narration in general, and esthetic effect are connected to one another. . . . Formulas are memory props and transition aids for the narrator. They are useful to him and comfortable, but they are additionally agreeable to him – just as the hearer is also delighted – when they turn up time and again, because he feels the organizing effect they have, and also simply because they are familiar to him. (1984, 44)

It is possible to add that in a situation in which an audience is very familiar with a range of formulae, the slightest variations in these will be perceived by the audience. This provides the opportunity for very subtle aesthetic effects based on such variation.

The research led by Propp and others into material which was unambiguously either oral in nature or closely related to oral material led, however, to an interest in formulaic elements in non-oral productions. The classic film Western, popular romantic fiction, and television soap operas certainly appeared to rely more heavily upon formulaic elements than did CANONICAL literature. Put crudely, we know what will happen on the last pages of a Barbara Cartland novel before we have started it, but even three-quarters of the way through *Wuthering Heights* on a first reading we will be in doubt as to its final conclusion. It is of course true that even canonical literature often contains formulaic elements which are invisible to modern READERS and which it is the function of research and scholarship to reveal to them. But beyond this it is apparent that the REPETITION of familiar elements, of recurrent patterns, plays a greater rôle in some literary WORKS than in others, and that it appears to be associated more with popular than with 'high' art.

John Cawelti has defined a formula as a 'combination or synthesis of a number of specific cultural CONVENTIONS with a more universal story form or archetype' (1977, 181; quoted in Yanarella & Sigelman 1988, 7). Cawelti finds a particular CULTURAL significance in such formulae because through many repetitions they become the conventional way in which particular images, symbols, MYTHS and THEMES are represented, and from which inferences concerning collective fantasies can be hazarded. This perhaps suggests that formulaic elements involve an IDEOLOGICAL element through an association with STEREOTYPES: the reader or TV viewer faced with a familiar formula is released from the need to confront a problem but merely swallows a preformed 'truth' without examining it. This seems to be true of certain formulaic elements in popular literature, but Max Lüthi argues, interestingly, that the formula's lack of individualization and concretization – what he calls its 'quasi-abstract generality' – actually leaves the reader's imagination more free to plump out the bare-bones of the formula as he or she wishes (1984, 21). (Compare Umberto Eco's argument in the entry for OPEN AND CLOSED TEXTS.)

This leads to the unusual idea that formulaic literature and art may actually allow for greater rather than lesser reader creativity. The negative side of such freedom, however, may be that the work offers fewer aesthetically productive constraints to the reader's fantasizing.

Fort / da From Freud's discussion in *Beyond the Pleasure Principle* (and also noted earlier in *The Interpretation of Dreams*) of the sound 'o – o – o – o' made by his grandson at the age of eighteen months. Freud interpreted the sound as the child's attempt to pronounce the German word 'fort' (gone) and claimed to have substantiated this when he witnessed the child throw a wooden reel on a piece of string over the end of his cot and repeat the sound, then pull it back into sight and repeat the word 'da' ('there'). For Freud this game represented the child's attempt symbolically to represent – and thus to gain symbolic mastery over – the unpleasant experience of his mother's regular separation from him. The game thus involved, for Freud, the symbolic representation of ABSENCE.

Jacques Lacan sees this game to manifest the *primordial symbolization* which inaugurates the conception of the signifying chain (1977, 215), and he claims that Freud's study of it exposed the origin of the REPETITION compulsion.

Frame 1. According to Mieke Bal, 'the space in which the CHARACTER is situated, or is precisely not situated, is regarded as the *frame*' (1985, 94).

2. In her book *Reading Frames in Modern Fiction* (1985) Mary Ann Caws applies the term *frame* to the phenomenon whereby many READERS find that certain passages in works of prose fiction 'stand out' from their surroundings. These passages are as it were *framed* by the surrounding TEXT, and this framing has important effects upon the manner in which they – and the WORK as a whole – are read. Caws suggests that such framing assumes an especial form in MODERNIST fiction as the idea of framing is called attention to or, we might say, FOREGROUNDED (1985, xi).

3. Following Erving Goffman's *Frame Analysis* (1974) the term is also used to denote various ways in which works of art (among other things) are aesthetically bounded and, thus, require or invite a range of different possible relationships with the art-consumer, with other works of art, or generally with extra-artistic reality.

A framed (or nested) NARRATIVE is either a 'narrative within a narrative', as in Henry James's *The Turn of the Screw*, or any narrative containing different narrative levels. With reference to such a work the term *frame narrator* refers to the NARRATOR of the outer narrative (alternatively, the term *outer narrator* is often used). Inner, or framed narratives are also known as *embedded narratives* or *Chinese Box narratives*, and where multiple embedding occurs it is sometimes known as *staircasing*.

4. Umberto Eco adopts definitions from Eugene Charniak and Michael Riffaterre in order to suggest a distinction between common frames – which are

the rules for practical life possessed by ordinary individuals – and INTERTEXTUAL frames, which are existing literary *topoi* or narrative schemes (1981, 21).

In P. N. Furbank's *Reflections on the Word 'Image'* there is an interesting discussion of what Furbank sees as the peculiarly modern tendency of 'abolishing the frame', a tendency he relates to a revolt against the idea of standards of REFERENCE and to an egalitarian desire to oppose the isolation of art in separated-off compartments (1970, 128–9).

5. In his *Framing the Sign*, Jonathan Culler explains why he prefers the term *frame* to the more conventional *context*. Most important, perhaps, is the argument that whereas the assumption is that a context is given, he avers that the truth is that 'context is not given but produced', and that

> The expression *framing the sign* has several advantages over *context*: it reminds us that framing is something we do; it hints of the frame-up . . ., a major use of context; and it eludes the incipient positivism of 'context' by alluding to the semiotic function of framing in art, where the frame is determining, setting off the object or event as art, and yet the frame itself may be nothing tangible, pure articulation. (1988, ix)

In his introductory comments to an essay by Barbara Johnson, Robert Young discusses briefly Jacques Derrida's use of the terms *parergon* (a term Derrida finds in Kant) and *ergon* as substitutes for *frame* and *work*:

> In the visual arts, the parergon will be the frame, or drapery, or enclosing column. The parergon could also be a (critical) text, which 'encloses' another text. (1981, 226)

He goes on, however, to note that Derrida's terms are connected with a more complex relationship than that of a simple inside/outside dichotomy, a complexity (we may add) which relates to Derrida's interest in the paradoxical nature of MARGINALITY.

According to Brian McHale, *frame-breaking* is characteristic of much POSTMODERNIST fiction: as, for instance, the statement of the narrator of John Fowles's *The French Lieutenant's Woman* that 'This story I am telling is all imagination. These characters I create never existed outside my own mind' (McHale 1987, 197). In like vein, John Frow argues that the task of the reader interested in coming to terms with a text's stored-up symbolic value who is also concerned with a reflexive integration of his or her situation, cannot be that of a 'correct' interpretation of the text:

> Rather than reproducing the text's official value, the reader must undertake a negative revalorization by 'unframing' it, appropriating it in such a way as to make it subversive of its own legitimacy and so *useful* in the class struggle. (1986, 228)

See also CLOSURE; EMBEDDING; FIGURE AND GROUND; OBSTINATION; SCRIPT (frame theory).

Free Direct Discourse See FREE INDIRECT DISCOURSE

Free Indirect Discourse. Also *Free Indirect Speech* or *Style*; *Narrated Monologue*; *Erlebte Rede*; *Style Indirecte Libre*; *Quasi-Direct Discourse* or *Substitutionary Narration*. In some usages these all represent the same general NARRATIVE technique, subdivisions within which are indicated by distinguishing between *Narrated Speech* and *Narrated Thought*, or between *Free Indirect Speech* and *Free Indirect Thought* (thus making the former phrase slightly ambiguous as it can either represent the umbrella term or a subdivision within it).

Usages do vary, however, and some commentators use *narrated monologue* to refer to a variety of Free Indirect Discourse in which there is indirect quotation of the words used in a CHARACTER's speech or thought. This would mean that non-verbalized thought-processes could not be represented by means of narrated monologue, and thus another term is called for: *psycho-narration*. According to Steven Cohan and Linda M. Shires, psycho-narration can be either *consonant* (following a character's own self-apprehension), or *dissonant* (moving back from a character's own perspective) (1988, 100).

The traditional way of defining what we can refer to as FID makes use of grammatical or linguistic evidence. This involves seeing FID as a midway point between Direct, and Indirect (or Reported) Discourse (DD and ID), or as a combination of the two which blends their grammatical characteristics in a distinctive mix. Thus Shlomith Rimmon-Kenan provides the following example, in which one can note that the third example retains the third person 'he' and past tense from ID, but in its truncation resembles the words in inverted commas in the DD example.

DD: He said, 'I love her'
ID: He said that he loved her
FID: He loved her (1983, 111)

FID often resembles ID minus the normal accompanying tag phrases (e.g. 'he suggested', 'she thought'). The standard grammatical/linguistic signs of FID are taken to be such things as DEICTICS referring to the character's own time or place (e.g. 'Tomorrow was Christmas'), the use of colloquialisms, etc. unlikely to have been used by the NARRATOR, abridgement such as is found in spoken but not, normally, written language, and the back-shift of tenses to be found in ID. When many of these characteristics are found together then a passage can be unambiguously FID, but FID may appear without any linguistic markers, such that only the semantic content of the passage in question can be adduced as evidence that one is dealing with FID.

Dorrit Cohn suggests that FID 'may be most succinctly defined as the technique for rendering a character's thought in their own idiom while maintaining the third-person reference and the basic tense of narration' (1978,

100). As already suggested, however, it is not just thought that can be represented, but also speech and, furthermore, attitudes, IDEOLOGICAL presuppositions, and so on. And the speech and thought can be either particular acts or ITERATIVE examples, or even thoughts that are potential in a character but unactualized. And, moreover, the thought may be either verbalized or unverbalized.

A matter of some contention has been that of the 'dual voice' hypothesis, with Pascal (1977) and Banfield (1982) taking up positions for and against the suggestion that FID involves the combination of two VOICES, those of the narrator and the character.

The general advantage gained from a use of FID is probably that of an apparently dramatic and intimately direct access to a character's thoughts or speech without the distracting presence of a narrator signalled by tag-phrases such as 'he thought' or 'she said'. By mixing FID, DD and ID an AUTHOR can achieve very considerable narrative flexibility.

A related term noted by Wales (1989, 77), who attributes it to Graham Hough, is *coloured narrative*. In this case, as Wales points out, the narrative is seen to be 'coloured' by the speech of a character, whereas in FID it is the speech which is coloured by the narrative voice.

Free Direct Discourse (or Speech or Thought or Style) is Direct Discourse represented without tags such as 'she thought' and 'he said'. Thus '"I'm tired", he said' would be Direct Speech, while 'I'm tired' (the inverted commas may be omitted altogether) is Free Direct Speech.

See also OBSTINATION.

Frequency Following Genette, the numerical relationship between events in a PLOT (or SJUŽET) and events in a STORY (or FABULA). This relationship can vary as follows:

i a singular event which is narrated once (*singulative narration*)

ii an event which occurs x times and is narrated x times (*multiple narration*)

iii an event which occurs once but is narrated more than once (*repetitive narration*)

iv a repeated event which is narrated only once (*iterative narration*)

A writer or NARRATOR's skill in varying frequency can play a crucial rôle in telling a STORY: stories which consist only of examples i and ii above can give a mechanical and unvarying impression, whereas the skilled use of REPETITION – especially forms of repetition with significant variation – can contribute to the achieving of an impression of depth and multiple PERSPECTIVE. Some writers (Genette draws particular attention to Proust) have made effective use of iterative narration to suggest underlying consistencies and patterns in a CHARACTER or situation.

Movement from frequencies i or ii to frequencies iii or iv in a narrative can be one of the means whereby an AUTHOR manipulates such things as DISTANCE, perspective, and dramatic involvement.

Genette uses the term *pseudo-iterative* to describe passages of narrative which claim to be iterative but which provide extended detail or include elements which by their very nature must be unique.

FTAs (face-threatening acts) See POLITENESS

Function Probably the most important usage in its relevance for Literary Studies is that to be found in Vladimir Propp's *Morphology of the Folktale*. Here Propp defines a narrative function as 'an act of a character, defined from the point of view of its significance for the course of the action' (1968, 21). Propp argued that although folk tales contain an extremely large number of different CHARACTERS, they contain a relatively small number of functions, functions which constitute the 'fundamental components' of the folk tale (1968, 21).

Propp's approach is STRUCTURALIST to the extent that it assumes a grammar of the folk tale such that functions play a rôle in an individual folk tale analogous to that played by parts of speech in a well-formed sentence. Thus just as the same word can perform different grammatical functions in different sentences (compare 'I set the table' with 'You win the first set'), so too can the same act perform a different narrative function in different folk tales (the appearance of a dragon could represent struggle in one tale, pursuit in another).

We find a similar distinction in the work of the RUSSIAN FORMALISTS between *function* and *device*. Of the two terms, device is the more neutral, and has a general reference independent of the context of a single TEXT. Thus Viktor Shlovsky writes of the device of eavesdropping in Dickens, which performs different functions in his works (Šklovskij 1971, 221). The distinction is perhaps clarified by reference to another term used by the Russian Formalists – that of *motivation*. According to Y. Tynyanov, 'Motivation in art is the justification of some single factor vis-à-vis all the others . . . Each factor is motivated by means of its connections with the remaining factors' (Tynjanov 1971, 130). (Contrast the definition of *motivation* given in the entry for ARBITRARY.) This resembles Barthes's comments, referred to below, on the *chaining* of sequences of actions to form functions. For the Russian Formalists, device and function were both involved in the concept of DEFAMILIARIZATION: a device could perform the function of defamiliarization at one time, but as it became familiar to readers it could lose this ability.

Propp limited the number of functions to 31, and argued that these appear in the individual folk tale in a sequence which is fixed and invariable – although, of course, not all functions appear in every tale. As with the governing of the choice and order of words in a sentence, the choice of alternative functions in a given STORY was seen to be governed by SYNTAGMATIC and PARADIGMATIC rules. However, arguments about just how many functions there are, about the rules governing their selection and use, and about which function(s) particular acts represent, soon began to suggest that even the relatively simple narratives with which Propp was concerned might resist the systematized analysis that he brought to bear on them. In general Propp's work has been more fruitful for the study of FORMULAIC LITERATURE than for more CANONICAL works, although it is clear that even in the canonical works of high literary CULTURES structural analysis on the basis of functions may have *some* validity.

In Claude Bremond's (1966 & 1973) development and modification of Propp's work, functions group together in threes to form sequences within which they punctuate three logical stages: possibility, process, outcome. For Bremond, each function opens a potentiality which can be ACTUALIZED or NON-ACTUALIZED.

In the work of Roland Barthes we find a further variation in the use of this term. In his 'Introduction to the Structural Analysis of Narrative' (1975), Barthes argues that a narrative can be seen as a large sentence and a sentence as, in some sense, a small narrative. He proceeds to isolate basic narrative units, and distinguishes between the *function* and the *index*. If the narrative units can be chained in sequences of actions they are termed functions, while if they perform a less structured rôle in the story they are termed indices. Jonathan Culler has suggested that Barthes's use of the term *function* in this essay is not a happy one, and that Barthes would have done better to have stuck to the term *lexie* (in the English translation: *lexia*) used in *S/Z* (Culler 1975, 202). In *S/Z* a lexia is a minimal unit of reading, a passage which has an isolable effect on the reader which can be distinguished from the effect of other passages (Barthes 1990, 13–14).

Gérard Genette uses the word in a different sense when he argues that 'there is no literary object strictly speaking, but only a *literary function*, which can invest or abandon any object of writing in turn' (1982, 4). Genette here uses *function* as equivalent to 'system-determined set of rules' in a STRUCTURALIST sense. Thus if one played chess with stones, one would be (according to this view) investing the stones with the chess function, a function which would subsequently abandon them if they were thrown away after the game.

A *kernel* function or type, according to some recent theorists of narrative, is one of the basic or essential components of a PLOT – or alternatively an element in a story that advances the action. (Compare HINGE, and see also *kernel event* in the entry for EVENT.)

See also FUNCTIONS OF LANGUAGE.

Functions of language In 1934, in his book *Language Theory*, Karl Bühler suggested an elegant way of classifying the different semantic functions performed by the linguistic SIGN. For him, it was illuminating to distinguish between the linguistic sign's *symbolic* function, which arose from its relation to things and states of affairs, its *symptomatic* function, which arose from its dependence upon its sender, whose inner states it expressed, and its *signalling* function, which rested upon its appeal to the listener, whose external or internal attitudes it directed much like a traffic sign. As Anders Pettersson (who refers to Bühler's distinctions) points out, Roman Jakobson's classification of the functions of language is 'a well-known further elaboration of Bühler's' (1990, 73).

Jakobson's rather more influential account is to be found in his essay 'Linguistics and Poetics', in which he claims that before the 'poetic function' is discussed, its place amongst the other functions of language must be defined. To do this, he argues, the constitutive factors in any act of verbal communication must first be surveyed. In a much-quoted passage he proceeds to do just this.

> The *addresser* sends a *message* to the *addressee*. To be operative the message requires a *context* referred to ('referent' in another, somewhat ambiguous, nomenclature), seizable by the addressee, and either verbal or capable of being verbalized; a *code* fully, or at least partially, common to the addresser and addressee . . .; and, finally a *contact*, a physical channel and psychological connection between the addresser and the addressee, enabling both of them to enter and stay in communication.
>
> (1960, 353; small capitals in original replaced by italics)

Jakobson then moves to claim that each of these six factors determines a different function of language. An orientation towards the context involves the *referential function*, which he sees as the leading task of numerous messages. The *emotive or expressive function* is focussed upon the addresser, and an example of the emotive function in a rare, pure form would be interjections such as 'Tut! tut!'. The *conative function* involves an orientation towards the addressee, typically in the vocative or imperative modes, and this function is distinguished by the fact that it is not liable to a truth test. The *phatic function* includes messages designed purely to 'keep the line open', to maintain communicative contact without actually communicating any information other than that needed to remain in such contact. The *metalingual function* involves checking that the same CODE is being used, where, for example, we ask a conversational partner to explain what he or she means by a particular word. Finally, a focussing on the message itself for its own sake, leads us to the *poetic function* of language. Robert Scholes has explicitly adapted the diagrammatic rendering of Jakobson's functions to describe the reading of a literary text, as below.

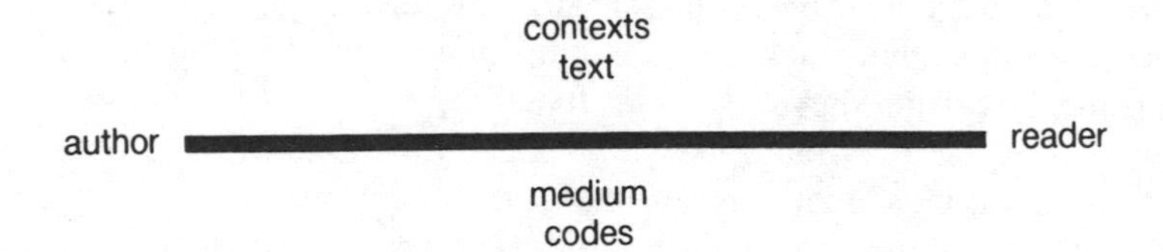

Jakobson's analysis has been influential but has not escaped criticism. Referring to the passage quoted above R. A. Sharpe makes the acid comment that banality does not preclude falsehood, and he goes on to argue that Jakobson's ignoring of the rôle played by interpretation in the literary arts is fundamental to what is wrong with his terminology (1984, 15).

See also the entry for the SHANNON & WEAVER MODEL OF COMMUNICATION, in which it is suggested that renderings of Jakobson's argument in diagrammatic form are much influenced by the Shannon and Weaver model, and also the entry for *S'ENTENDRE PARLER*.

G

Gap See ABSENCE; ELLIPSIS; FIGURE; PHENOMENOLOGY

Gatekeeping In some ways comparable to AGENDA SETTING, the concept of gatekeeping comes from media studies and, more particularly, studies of the attitude-forming effects of the news media. Just as a literal gatekeeper will prevent you getting through the door to talk to an important person, metaphorical gatekeepers prevent certain news items, or opinions, or interpretations from reaching a larger public. It can be argued that the formation and defence of a CANON can have a gatekeeping function, and that gatekeepers work with certain IDEOLOGICAL presuppositions that help them to carry out their function – as much in university departments of literature as on TV news desks.

Gaze The concept of the gaze constitutes an important if opaque element in Jacques Lacan's theories concerning the formation of subjectivity. In his *The Four Fundamental Concepts of Psychoanalysis* he states that

> The gaze is presented to us only in the form of a strange contingency, symbolic of what we find on the horizon, as the thrust of our experience, namely, the lack that constitutes castration anxiety.
>
> The eye and the gaze – this is for us the split in which the drive is manifested at the level of the scopic field.
>
> . . .
>
> In our relation to things, in so far as this relation is constituted by the way of vision, and ordered in the figures of representation, something slips, passes, is transmitted, from stage to stage, and is always to some degree eluded in it – that is what we call the gaze. (1979, 72–3)

Elsewhere, Lacan subjects Sartre's view of the gaze to critical scrutiny (1979, 84, 89, 182), insisting (if I understand aright) on the necessarily reflexive nature of his own conception of the gaze.

The concept has been more utilized in film than in literary criticism, and especially by FEMINIST critics who have used it to explore the element of desired (but perhaps denied) power in the voyeuristic male utilization of the gaze in the cinema. Various feminist critics have pointed out that in 'classic' Hollywood films the gaze is very much in the possession of men, thus depriving women of power and of significant subjectivity. Women are to be looked at, not to look; their alienation from the gaze is an aspect of their passivization and REIFICATION.

See Mulvey (1985) and Doane (1988) for further discussion of this topic.

Gender In current FEMINIST usage, gender is defined as characteristics of socio-cultural origin attributed to the different biological sexes. Within Linguistics this usage is sometimes varied in order to avoid confusion with linguistic gender, but generally speaking feminist influence has succeeded in establishing that *gender* involves society and or culture and *sex* involves biology.

A *genderlect* is a term used to describe linguistic characteristics which in a given society or CULTURE are specific to members of one gender.

See also IDIOLECT; SOCIOLECT.

Geneva School See PHENOMENOLOGY

Genotext and phenotext A distinction coined by Julia Kristeva and popularized in English as a result of its use by Roland Barthes, among others. According to Kristeva, the phenotext is 'the verbal phenomenon as it presents itself in the structure of the concrete statement', while the genotext

> 'sets out the grounds for the logical operations proper to the constitution of the subject of the enunciation'; it is 'the place of structuration of the phenotext'; it is a heterogenous domain: at the same time verbal and of the nature of drives ('pulsionnel') (it is the domain 'where signs are cathected by drives').
> (Barthes 1981b, 38, quoting from an interview with Kristeva published in 1972)

As can be seen, the distinction is not an easy one to gloss, but a key element seems to reside in the final quotation given by Barthes: whereas the phenotext is the purely verbal element contained in and constituting a particular UTTERANCE, the genotext is that complex of genetic forces where the non-verbal meshes with and expresses itself in the verbal.

The distinction has enjoyed a limited circulation amongst literary critics engaged in theorizing literary genesis.

Glissement See SLIPPAGE

God trick See CENTRE

Gram See DIFFÉRANCE

Grammatology In *Of Grammatology* Jacques Derrida attributes this term to Littré, from whom he quotes: 'A treatise upon Letters, upon the alphabet, syllabation, reading, and writing.' Derrida notes that to his knowledge the word has been used in the present century only by I. J. Gelb in *A Study of Writing: The Foundations of Grammatology* (1952), which book, Derrida claims, 'follows the classical model of histories of writing' (Derrida 1976, 323 n4). Derrida uses the word to indicate a 'science of writing' that is, he believes, showing signs of liberation all over the world (1976, 4). This should not, however, be understood in terms of a full-formed science the basic principles of which are established and final, but – he implies – a science in which everything is questioned, including its own basis and history (1976, 28). For Derrida's use of the word ÉCRITURE see the separate entry.

In 'Linguistics and Grammatology' Derrida suggests that the word grammatology should replace the word semiology in the programme set out in Ferdinand de Saussure's *Course in General Linguistics*, as this will give the theory of writing the scope needed to counter LOGOCENTRIC repression and the subordination to linguistics.

Grand narratives and little narratives From the French terms *grand récit* and *petit récit*. The terms have been given widespread currency by Jean-François Lyotard's book *The Postmodern Condition: A Report on Knowledge* (1984; first published in French 1979). Lyotard distinguishes the modern by its association with what he calls grand narratives:

> I will use the term *modern* to designate any science that legitimates itself with reference to a metadiscourse of this kind making an explicit appeal to some grand narrative, such as the dialectics of Spirit, the hermeneutics of meaning, the emancipation of the rational or working subject, or the creation of wealth.
> (1984, xxiii)

A grand narrative, then, is a means for EMPLOTTING a life or a CULTURE; to give one's actions or one's life meaning one pictures oneself as a CHARACTER within an already-written narrative whose final conclusion is assured in advance.

In popular usage the term is typically applied to IDEOLOGICAL systems such as MARXISM or religious outlooks such as Christianity or Islam. In Lyotard's view the day of such all-embracing, totalizing systems of belief has passed. From now on we have to derive meaning from little narratives, from local justifications:

> The narrative function is losing its functors, its great hero, its great dangers, its great voyages, its great goal. It is being dispersed in clouds of narrative language elements – narrative, but also denotative, prescriptive, and so on. (1984, xxiv)

To describe this great shift from the time of the grand narrative to that of the little narrative, Lyotard argues that the society of the future falls less within the province of a Newtonian anthropology such as structuralism or systems theory, and more with a pragmatics of language particles, a world of many language games (1984, xxiv).

Ground See FIGURE AND GROUND

Gyandry See ANDROGYNY

Gynocentric See ANDROCENTRIC

Gynocratic / gynecocratic That which is ruled by women. Thus a gynocratic society would be one in which women held power, in contrast to an ANDROCRATIC society, in which power would be (and has been) in the hands of men.

Gynocritics According to Elaine Showalter the term gynocritics is an invention of hers to describe that FEMINIST criticism which studies women *as writers*, 'and its subjects are the history, styles, themes, genres, and structures of writing by women; the psychodynamics of female creativity; the trajectory of the individual or collective female career; and the evolution and laws of a female literary tradition' (1986, 248).

See also ÉCRITURE; ÉCRITURE FÉMININE; FEMINISM.

H

Hegemony A term used by MARXISTS to describe the maintenance of power without the use, or direct threat, of physical force; normally by a minority class whose interests are contrary to those over whom power is exercised. Its modern use stems from the work of the Italian communist Antonio Gramsci, whose most influential work was written while he was incarcerated by Mussolini's fascists, and published in various collections after the Second World War. Although Gramsci recognized the importance of the use of force by Italian fascism, he was perhaps more preoccupied with what he saw as the '"spontaneous" consent given by the great masses of the population to the general direction imposed on social life by the dominant fundamental group; this consent is "historically" caused by the prestige (and consequent confidence) which the dominant group enjoys because of its position and function in the world of production' (Gramsci 1971, 12).

Helper See ACT / ACTOR

Hermeneutic code See CODE

Hermeneutics Within Anglo-American literary-critical circles the term hermeneutics has often been used as a loose synonym for interpretation, and indeed its primary reference is to the understanding of texts and the understanding of understanding – especially with relation to the discipline of theology.

More recently, however, use of the term has generally implied specific reference to the German hermeneutic tradition, a tradition which has become better known in Britain and America during the past couple of decades through the work of E. D. Hirsch and Wolfgang Iser. One of the reasons why the German hermeneutic tradition was generally neglected in the 1970s and 1980s within Anglo-American theory can be traced back to the importance of a historical perspective to this tradition, a perspective which accorded ill with such synchronic and anti-historicist approaches as STRUCTURALISM and (to a certain extent) DECONSTRUCTION.

The founder-fathers of the German hermeneutic tradition are Friedrich Schleiermacher (1768–1834) and Wilhelm Dilthey (1833–1911), although both built on the work of German Protestant theologians of the seventeenth century whose development of methods of Biblical interpretation had a clearly theological imperative. It is to Schleiermacher that we owe the concept of the hermeneutical circle (see below), although it was Dilthey who gave it its name.

Three other very important names in the German hermeneutic tradition are those of Edmund Husserl (1859–1938), Martin Heidegger (1889–1976), and Hans-Georg Gadamer. Husserl's development of PHENOMENOLOGY had a number of very important implications for theories of interpretation, especially in terms of the active and 'completing' rôle that the act of consciousness has in CONCRETIZING its incompletely perceived objects of perception. Heidegger's main contribution to hermeneutical theory is, arguably, his anti-individualistic and historicist view of the process of interpretation.

The word hermeneutics is etymologically related to the name of the messenger-God Hermes, who, as Richard Palmer points out in his book *Hermeneutics*, is, significantly, 'associated with the function of transmuting what is beyond human understanding into a form that human intelligence can grasp' (1969, 13). Palmer points out that hermeneutics has traditionally had two main focuses of concern: 'the question of what is involved in the event of understanding a text, and the question of what understanding itself is, in its most foundational and "existential" sense' (1969, 10). He thus argues for the existence of *two* hermeneutic traditions:

> There is the tradition of Schleiermacher and Dilthey, whose adherents look to hermeneutics as a general body of methodological principles which underlie interpretation. And there are the followers of Heidegger, who see hermeneutics as a philosophical exploration of the character and requisite conditions for all understanding.
>
> . . .

> Gadamer, following Heidegger, orients his thinking to the more philosophical question of what understanding itself is; he argues with equal conviction that understanding is an historical act and as such is always connected to the present. (1969, 46)

Perhaps the aspect of the hermeneutic tradition which is best-known in Anglo-American circles is that rendered by the term 'the hermeneutic circle'. The term is used to express the seeming paradox that the whole can be understood only through an understanding of its parts, while these same parts can be understood only through an understanding of the whole to which they belong.

E. D. Hirsch is the theorist who has done most to bring the work of the German hermeneuticists to the attention of Anglo-American readers. Indeed, in his essay 'Objective Interpretation', (first published in 1960 and reprinted in *Validity in Interpretation* [1967]), Hirsch states that 'my whole argument may be regarded as an attempt to ground some of Dilthey's hermeneutic principles in Husserl's epistemology and Saussure's linguistics' (1967, 242, n30). In his subsequent book, *The Aims of Interpretation* (1976), Hirsch describes his relationship to the German hermeneutic tradition in more detail.

It is for Gadamer that Hirsch reserves his strongest criticisms, focussing on his *Truth and Method* in an essay entitled 'Gadamer's Theory of Interpretation', first published in 1965 and reprinted as an appendix to *Validity in Interpretation. Truth and Method* was first published as *Wahrheit und Methode* in Tübingen 1960; the English translation (1975, revised 1989) is of the second edition, published Tübingen 1972. Hirsch is extremely critical of Gadamer's argument that the author's perspective and the interpreter's perspective must be fused in a fusion of horizons (*Horizontverschmelzung*), asking, 'How can an interpreter fuse two perspectives – his own and that of the text – unless he has somehow appropriated the original perspective and amalgamated it with his own?' (1967, 254).

Hirsch's own distinction between MEANING AND SIGNIFICANCE has been seen by more than one commentator as an attempt to break the then-current orthodoxies of New Critical anti-contextualism by going back to the origins of the German hermeneutic tradition – in particular, to the work of Friedrich Schleiermacher. Schleiermacher believed that the process of understanding reversed the process of composition, for instead of starting with the AUTHOR's mental life and proceeding to textual embodiment or projection it started with the text and worked its way back to its originating mental life. It is easy to see how Hirsch's insistence on equating *meaning* with authorial intention builds on such a view.

If the historicist emphases of the German hermeneutic tradition militated against its catching the interest of many Anglo-American theorists during the heyday of structuralism and deconstruction, two new developments may help to change this situation. The first is the rise of the NEW HISTORICISM, which is helping to make the historicist emphases of the hermeneutic tradition much less unfashionable, and the second is the fact that the popularity of a Bakhtinian emphasis on the DIALOGIC may, similarly, make some of the arguments of Hans-Georg Gadamer sound a little more relevant to Anglo-Saxon ears.

Other theorists who have been particularly influenced by the German Hermeneutic tradition are Roman Ingarden, Paul Ricoeur, the members of the Geneva School of Criticism (see the entry for PHENOMENOLOGY) and those of the German RECEPTION THEORY school.

See also the entry for HORIZON. For 'hermeneutics of suspicion', see OPPOSITIONAL READING.

Heterodiegesis See DIEGESIS AND MIMESIS

Heteroglossia In the writing of Mikhail Bakhtin, that multiplicity of social voices linked and interrelated DIALOGICALLY which enters the novel through the interplay between authorial speech, NARRATOR speech, 'inserted genres', and CHARACTER speech (1981, 263).

The glossary provided in Bakhtin's *The Dialogic Imagination* notes that heteroglossia is determined contextually and extra-linguistically as well as intra-linguistically: 'all utterances are heteroglot in that they are functions of a matrix of forces practically impossible to recoup' (1981, 428).

In Bakhtin's usage, according to the same source, *polyglossia* refers more specifically to the co-existence of different national languages within a single CULTURE.

See the longer entries for DIALOGIC and POLYPHONY.

Heteronomous objects See CONCRETIZATION

Heterosexism See HOMOPHOBIA

Heuristic reading See MEANING AND SIGNIFICANCE

Hinge 1. Mieke Bal uses the term *hinge* in the context of a discussion of FOCALIZATION, suggesting that it can be applied to passages of NARRATIVE with either a double or an ambiguous focalization.

2. In the writing of Jacques Derrida *brisures* are often rendered as *hinge-words* in English, although Maud Ellmann has suggested the alternative translation of *cleavage* (Ellmann 1981, 192). These hinge-words contain, following Derrida, a paradoxical logic which must be explored by deconstructive analysis. Commenting upon Derrida's use of this term, Robert Young suggests that the effect of such hinge-words 'is to break down the oppositions by which we are accustomed to think and which ensure the survival of metaphysics in our thinking' (1981, 18).

3. What Roland Barthes refers to as *nuclei* are sometimes described as *hinge-points* in English: see the entry for EVENT.

4. Clemens Lugowski uses the term to refer to a 'crucial point in the action' of a narrative (1990, 56).

Historicism See NEW HISTORICISM

Hommelette A coinage of Jacques Lacan's which combines the senses of 'little man' and omelette, and by which he seeks to describe the pre-Oedipal psychic condition of the child. The child is a little man (the GENDER bias is impossible to avoid) in as much as it contains *in posse* the whole of the later adult, but in diffused and undistinguished form (the child has no clear sense of self distinct from non-self). Lacan refers to the term as a 'joky' alternative to the more technical 'lamella' (1979, 197).

Homodiegetic See DIEGESIS AND MIMESIS

Homology Also *isomorphism* or *structural parallelism.* A correspondence or similarity which establishes a significant pattern or structural REPETITION. This can either be within a given literary WORK, or (in STRUCTURALIST theory) between the STRUCTURE of a language and the structure of, for instance, the human UNCONSCIOUS. Structuralists have also argued for the existence of homologies between the language system and other systems, from kinship relationships to literature, claiming that literature (seen as a total system of relations) is itself structured like a language.

Similarly, in NARRATIVE theory, structuralist theorists have argued that there is a homology between the syntax of a grammatical sentence and the larger narrative 'syntax' of a literary work, and have used terms from linguistics to describe particular narrative functions (e.g. MODE). Tzvetan Todorov has suggested that we think of a literary CHARACTER as a noun and an action as a verb, and their combination as the first step towards narrative (1969, 84). (See the entry for LINGUISTIC PARADIGM.)

The Romanian-French MARXIST critic Lucien Goldmann has made considerable use of the concept in his work, suggesting homologies between class situation, world view, and artistic form.

Fredric Jameson has implied that homology should be distinguished from MEDIATION, for whereas homology involves resemblance at a structural level, mediation involves a connective relationship which has an element of dependence or causality between the different elements linked by mediation (1981, 43). Usage tends to be less tidy than this, however, and although Jameson's suggested distinction makes good sense it does not reflect the way the terms are actually used.

Homonymy Following Brian McHale, the reappearance of an entity from one fictional world in another, but with essential changes. The extent of these changes is important: to count as a case of homonymy there must be variants in essential properties – the main CHARACTERS of Samuel Richardson's *Pamela* and of Henry Fielding's parodic *Shamela*, for example.

Where the changes involve only accidental qualities, then we have a case of quasi-homonymy. And where a character is arguably unchanged, as with the character Cordelia in Shakespeare's *King Lear* and Nahum Tate's rewritten version of the same play, then we have a case of *transworld identity* (McHale 1987, 35–6).

Homophobia Fear (and hatred) of homosexuals. In his discussion of the term in Wright (1992) Jeffrey Weeks notes that the popularization of the concept is generally attributed to George Weinberg (1972), who, according to Weeks, argued that 'the real problem was not homosexuality but society's reaction to it' (Wright 1992, 155).

The term has been important in literary-critical discussion of a range of writers. Eve Kosovsky Sedgwick has commented on the term that the word is etymologically nonsense, presumably because it suggests fear of human beings rather than of homosexuals. She adds that

> A more serious problem is that the linking of fear and hatred in the '-phobia' suffix, and the word's usage, does tend to prejudge the question of the cause of homosexual oppression: it is attributed to fear, as opposed to (for example) a desire for power, privilege, or material goods. (1993, 219, n1)

She notes that the term *heterosexism* offers a possible alternative, and since 1985, when Sedgwick's book was first published, use of the term *heterosexism* has become more widespread.

See also HOMOSOCIAL; SEXISM.

Homosocial A term adopted and popularized by Eve Kosofsky Sedgwick in her book *Between Men: English Literature and Male Homosocial Desire* (1993, first published 1985). According to Sedgwick,

> 'Homosocial' is a word occasionally used in history and the social sciences, where it describes social bonds between persons of the same sex; it is a neologism, obviously formed by analogy with 'homosexual,' and just as obviously meant to be distinguished from 'homosexual.' In fact, it is applied to such activities as 'male bonding,' which may, as in our society, be characterized by intense homophobia, fear and hatred of homosexuality. (1993, 1)

Horizon Members of the group around Mikhail Bakhtin (P. N. Medvedev, V. N. Vološinov, and Bakhtin himself) regularly used this term to suggest the borders of possibility (normally) constraining a READER. Thus we come across terms such as *ideological horizon, socio-linguistic horizon*, and *axiological horizon* (i.e. the limit of possible evaluative acts) in their work. The usage has something in common with theories concerning the manner in which IDEOLOGICAL situations restrict interpret[at]ive possibility, and also with Michel Foucault's view of the way in which the ARCHIVE of possible DISCOURSES limits an individual's access to knowledge. See the discussions in Todorov (1984).

The entry for HERMENEUTICS should be consulted for Hans-Georg Gadamer's theory of the 'fusion of horizons', and for 'horizon of expectations' see the entry for RECEPTION THEORY.

Hot and cool media According to Marshall McLuhan, there is a basic principle which distinguishes hot media such as radio and film from cool media such as the telephone and TV.

> A hot medium is one that extends one single sense in 'high definition.' . . . Telephone is a cool medium, or one of low definition, because the ear is given a meager amount of information. And speech is a cool medium of low definition, because so little is given and so much has to be filled in by the listener.
>
> (1964, 22-3)

Furthermore, McLuhan argues, hot media do not leave so much to be filled in or completed by the audience, whereas cool media require the receiver of the information to fill in much. Hot media, therefore, are low-participation media, whilst cool media are high-participation media.

McLuhan's assimilation of technology and message raises a number of problems here – problems which have led to his being accused of TECHNOLOGICAL DETERMINISM by many commentators. For clearly although the telephone gives a meagre amount of information compared to a film *in a purely technical sense*, the information given in a telephone conversation can be semantically very rich indeed.

Compare READERLY AND WRITERLY TEXTS, noting the difference that emphasis on a medium and emphasis on a TEXT makes.

Humanism This is a term that has undergone a marked change of fortunes during the last two or three decades. Michèle Barrett has pointed out that this is particularly true in the field of culture, 'to the extent that in some circles *it is assumed that* "humanist" is a derogatory term' (1991, 93).

The current attribution of a peculiarly pejorative force to this term within Anglo-American (and especially Anglo) literary theory can probably be dated from the publication of Louis Althusser's *For Marx* (1969; first published in French 1966). Althusser declares at the beginning of this work that his aim is to oppose 'Marxist humanism' and the '"humanist" interpretation of Marx's work' (1969, 10). In his essay 'Marxism and Humanism' (1963; included in *For Marx*), Althusser makes clear that a key opposition for him is that between the insistence on the end of *class* exploitation (the Marxist goal), and the attainment of human freedom (which, he implies, is the goal of humanism) (1969, 221). Althusser locates a non-Marxist humanism in the works of the early Marx, in which, according to Althusser, 'the proletariat in its "alienation" represents the human essence itself, whose "realization" is to be assured by the revolution' (1969, 221–2). For Althusser this positing of a human essence lies at the heart of humanism, and many subsequent theorists have accepted his emphasis. For them such a posited human essence is beyond history and beyond society, and thus is essentially idealist rather than REALIST, often involving the projection of the characteristics of one form of society on to human beings at large (thus one often finds reference to *bourgeois humanism* or *liberal humanism*). Furthermore, (the argument continues), humanism typically situates the human essence in individual human beings rather than in social structures or CULTURAL formations: humanism is thus idealistic, ahistorical, and individualistic.

Althusser's view of the early Marx is not uncontroversial. In the early *Economic and Philosophical Manuscripts of 1844* Marx does, it is true, refer to 'man's' (by which he means the human individual's) '*essential* being', but he also says that this is what man makes from his 'life activity'. He also argues that man is a being that treats its species as its own essential being – that treats itself 'as a species being' (Marx 1970b, 113). None of these arguments, it will be noted, posit an essence that is outside of history, and indeed Pauline Johnson has gathered evidence to suggest that Marx's theory of a human essence does not imply any 'natural' human attributes, but 'refers specifically to the process of transformation and development which characterizes the history of the species' (1984, 36). Johnson claims that this is certainly how both Lukács and Adorno interpreted Marx's 'human essence' (1984, 100).

Michèle Barrett argues that to assume that the term can and should be dispensed with,

> is historically a great injustice, in that it ignores the immensely progressive role that humanism – as an 'ideology' – has played. In particular, one can point to the honourable tradition of humanism as a secularising force and, indeed, to its enormously important role in contemporary politics. (1991, 93)

Hybrid According to Mikhail Bakhtin a hybrid utterance is one within which two different linguistic consciousnesses co-exist. Bakhtin analyses some parodic/ironic passages in Charles Dickens's *Little Dorrit* as examples of hybrid utterances (1981, 302–7; see also FREE INDIRECT DISCOURSE).

In more recent use, a *hybrid text* can be one formed by cutting two other texts together – in either a planned or a random manner. The term hybrid text can also be used to describe a text in which two separate, and often opposed, elements can be detected, on a thematic or an IDEOLOGICAL level.

Hypodiegetic See DIEGESIS AND MIMESIS

Hypostatization See REIFICATION

Hypothesis driven See SOLUTION FROM ABOVE / BELOW

I

Id See TOPOGRAPHICAL MODEL OF THE MIND

Ideal reader See READERS AND READING

Ideogram See DEFORMATION

Ideologeme Following Fredric Jameson, 'the smallest intelligible unit of the essentially antagonistic collective discourses of social classes' (1981, 76). Formed by analogy with such terms as *phoneme* (the smallest intelligible unit of significant sound in a language). Compare SEMEME, STYLEME, and similar coinages.

See also the more detailed entry for IDEOLOGY.

Ideological horizon See HORIZON

Ideological State Apparatuses (ISAs) See IDEOLOGY

Ideology No definition of this term can hope to provide a single and unambiguous meaning; instead a cluster of related but not always compatible meanings have to be indicated. What nearly all commentators agree upon is that the present-day use of the term refers to a *system of ideas*: according to some usages an ideology may include contradictory elements, but if so these elements are somehow brought into a functioning relationship which obscures these contradictions for the person or people by whom the ideology is lived. An ideology is thus a way of looking at and interpreting – of 'living' – the world. A further point of agreement is that ideologies are *collectively held*; a purely personal system of ideas would not normally be called an ideology.

Alternative present-day usages of the term differ on a number of other points, however: whether an ideology is necessarily a form of false consciousness or whether it can give a true and objective insight into reality (probably the most important area of disagreement or variation of usage); and whether an ideology represents the interests of a particular social class – and if so, how. In his recent book *Ideology: An Introduction*, Terry Eagleton suggests six broad definitions of the term: (i) 'the general material process of production of ideas, beliefs and values in social life'; (ii) 'ideas and beliefs (whether true or false) which symbolize the conditions and life-experiences of a specific, socially significant group or class'; (iii) 'the *promotion* and *legitimation* of the interests of such social groups in the face of opposing interests'; (iv) such promotion and legitimation when carried out by a 'dominant social power'; (v) 'ideas and beliefs which help to legitimate the interests of a ruling group or class specifically by distortion and dissimulation'; (vi) similar false and deceptive beliefs which arise 'not from the interests of a dominant class but from the material structure of society as a whole' (1991, 28–30).

For the French Marxist philosopher Louis Althusser, class societies are maintained as much by a consensus produced ideologically, through what he calls Ideological State Apparatuses (ISAs), as by repression through Repressive State Apparatuses (RSAs). The ISAs include the educational ISA, the family ISA, the legal ISA, the political ISA, the trade-union ISA, the communications ISA and the cultural ISA. Althusser's short definition of ideology is that it is, 'a "representation" of the imaginary relationship of individuals to their real conditions of existence' (1971, 152); for him ideology is always a distortion of reality, in contrast to science, which is not.

It is worth comparing Althusser's ISA with the way in which Foucault defines a DISCOURSE, and indeed Terry Eagleton has pointed out that Foucault and his followers effectively abandon the concept of ideology altogether and replace it with 'the more capacious "discourse"'. Eagleton notes, however, that such a substitution relinquishes a useful distinction, as '[t]he force of the term ideology lies in its capacity to discriminate between those power struggles which are somehow central to a whole form of social life, and those that are not' (1991, 8).

Let us pause at this moment to ask what all this has – or might be argued to have – to do with literature. We can start off by observing that Althusser's list of ISAs is very reminiscent of Marxist descriptions of the SUPERSTRUCTURE, and that literature has been categorized as superstructural by certain Marxists. From this perspective, to say that Marx sees literary works as part of 'a larger ideological superstructure' (Forgacs 1986, 170) might suggest that Marx held literature to be entrapped by its function of disseminating the ideas of the current ruling class – and this clearly REDUCTIVE view has found its adherents (although not, it should be pointed out, David Forgacs himself).

A more optimistic application of these ideas to literature would proceed along the lines recommended by Marx and Engels in *The German Ideology*. Rather than taking literary works at their own estimation these need to be read in the light of 'the contradictions of material life' as lived by the author and his or her contemporaries. This is probably the most important and representative element in the various forms of Marxist literary criticism that have developed in the present century and is the Marxist version of genetic criticism.

See also INCORPORATION.

Idiolect A term used by linguisticians to describe the features of a particular person's language which mark out him or her *individually* from others. An idiolect is thus distinguished from a *dialect*, which refers to the language characteristics marking out a *community* (geographical, social, educational) from others. Linguisticians normally restrict the primary reference of both terms to speech, but they are also applied by extension to written language as well.

Success in giving a particular literary CHARACTER a distinctive idiolect can be an important aspect of a writer's success in characterization. Linguistic analysis of Jane Austen's fiction has confirmed that one of the reasons why readers find her characters to be possessed of such convincing independent life is that she is so adept at granting them distinctive idiolects.

Compare *genderlect* in the entry for GENDER; and SOCIOLECT.

Illocutionary act See SPEECH ACT THEORY

Imaginary / symbolic / real A tripartite distinction found in the works of Jacques Lacan. Michèle Barrett gives an elegant short account of the force these three terms have in Lacan's writings:

> The 'imaginary order' includes images and fantasies, both conscious and unconscious; it is a key register of the ego and its identifications, evolving from the mirror stage but continuing in adult relationships; it particularly includes material from pre-verbal experience. The 'symbolic order', on the other hand, is the domain of symbolisation and language, and it is through the social and cultural processes of this symbolic order that the subject can represent desire and thus be constituted. The 'real' is defined as that which exists outside symbolisation, and outside the analytic experience which is necessarily contained by the limits of speech: it is that which is formally outside the subject. (1991, 102; Barrett notes that she bases her definition on Lacan's translator's [Alan Sheridan's] note [see below] and on Benvenuto & Kennedy 1986, 80–82)

For a more extended discussion of these three 'orders', see the separate entries in Wright (1992). See also Vincent Crapanzano's 'The Self, the Third, and desire', chapter 3 of Crapanzano (1992). Lacan's translator, Alan Sheridan, also includes a useful commentary on these linked terms in a note at the end of Lacan's *The Four Fundamental Concepts of Psychoanalysis* (1979, 279–80).

Immasculation According to Judith Fetterley, '[t]hough one of the most persistent of literary stereotypes is the castrating bitch, the cultural reality is not the emasculation of men by women, but the *immasculation* of women by men' (1978, xx). By immasculation Fetterley means that process whereby 'as readers and teachers and scholars, women are taught to think as men, to identify with a male point of view, and to accept as normal and legitimate a male system of values, one of whose central principles is misogyny' (1978, xx). Thus the title of her book, *The Resisting Reader: A Feminist Approach to American Fiction* (1978) proposes such resistance on the part of the reader as a necessary counter to the pressure of immasculation.

Implicature (conversational) See SPEECH ACT THEORY

Implied author See AUTHOR

Implied reader See READERS AND READING

Incorporation A term from MARXIST political discussion referring to the manner whereby opposition is neutralized by being incorporated into the dominant structures of power. Thus – the argument runs – the freedom enjoyed by speakers at Speakers' Corner in London is often paraded as evidence of the existence of free speech in Britain, whereas the relative ineffectiveness of speaking to a small gathering of persons in the age of the mass media suggests that this is a form of incorporation: the free speech is of more use in advertising the benefits of British democracy than it is in actually achieving anything that the free speakers want. Incorporation is, then, a form of 'If you can't beat them, get them to join you'.

A number of the CULTURAL MATERIALISTS have explored the concept and the issues it raises, although not always using the term 'incorporation'. Alan Sinfield's preferred terms are 'entrapment and containment'. Sinfield credits NEW HISTORICISTS with having developed the 'entrapment model' of IDEOLOGY and power, 'whereby even, or especially, maneuvers that seemed designed to challenge the system help to maintain it' (1992, 39), although he notes that entrapment is a concept found also in functionalism, structuralism and Althusserian Marxism.

Containment, of course, is what an adversary who knows that victory is impossible seeks to achieve to avoid defeat. If you can't put out the forest fire then you try to contain it.

Yet another related term is *recuperation*; this is a term used to describe a strategy whereby controlling authorities concede certain ideological positions to oppositional forces, but only so as to be able to incorporate these in a larger system of beliefs which reflects the interests of the controlling authorities.

The term *appropriation* has a similar force to incorporation, but the implied 'take-over' is more complete. Whereas an argument which is incorporated may retain some of its integrity, one which is appropriated becomes the total property of the appropriating authority.

Indeterminacy See ELLIPSIS

Influence See REVISIONISM

Inscribed reader See READERS AND READING

Instance In his essay 'Ideology and Ideological State Apparatuses' Althusser refers to the revolutionary character of Marx's conception of the 'social whole':

> Marx conceived the structure of every society as constituted by 'levels' or 'instances' articulated by a specific determination: the *infrastructure*, or economic base (the 'unity' of the productive forces and the relations of production) and the *superstructure*, which itself contains two 'levels' or 'instances': the politico-legal

(law and the State) and ideology (the different ideologics, religious, ethical, legal, political, etc.). (Althusser 1971, 129)

Thus Althusser divides up all *social formations* into three 'instances': the IDEOLOGICAL, the economic and the political. The term is often used by Althusser in connection with Marx's reference to the determination in the last instance by the (economic) mode of production (cited in Althusser 1969, 111). It should be noted that this is a somewhat different use of the word 'instance'.

Althusser also uses the term *determinant instance* – in other words, that force (ideological, economic, or political) that is DOMINANT and decisive in a particular CONJUNCTURE.

Intellectuals The Italian MARXIST Antonio Gramsci makes a useful distinction between *traditional* and *organic* intellectuals: according to him an organic intellectual is one who remains a member of and committed to his or her class of origin – a working-class leader, for example – whereas a peasant who leaves his class to become a Jesuit priest becomes a traditional intellectual because such a move takes him out of his class of origin. See the chapter entitled 'The Intellectuals' in Gramsci (1971).

The distinction is relevant to arguments about working-class writing, especially with regard to the question as to whether a member of the working class who becomes a writer can remain an organic intellectual, can claim the same relationship to the working class as, say, an industrial worker, or whether becoming a writer necessarily ALIENATES the individual from his or her class of origin.

Intended reader See READERS AND READING

Intercalated dialogue See INTERIOR DIALOGUE

Interior dialogue Alternatively (in Mikhail Bakhtin's usage) *internal dialogue* or *microdialogue*. A DIALOGUE between two well-defined voices within the single consciousness of a literary CHARACTER (or, in a wider usage, of a real human being), and the NARRATIVE representation of this process.

Interior dialogue involves more than the representation of a character's verbalized thought-processes in which questions are asked and answered. To count as genuine interior dialogue the questions and answers must stem from two voices which represent different and as it were *personified* attitudes, beliefs, or characteristics. A good example occurs towards the beginning of the tenth chapter of Charlotte Brontë's *Jane Eyre*, in which we have represented a long dialogue between different aspects of Jane Eyre's personality or identity. Mikhail Bakhtin gives another example from Book 11 of Part 4 of Feodor Dostoevsky's *The Brothers Karamazov* (Bakhtin 1984, 255).

Vincent Crapanzano provides a comparable concept with his term *shadow dialogue*. According to him the term refers to

> those dialogues that one partner to the primary dialogue has with an interlocutor, real or imaginary, who is not present at the primary dialogue. Such dialogues are 'silent,' 'mental,' 'quasi-articulate,' 'beneath consciousness' though capable, at least in part, of becoming conscious. They are analogous to thought when it is conceived as a conversation. (Crapanzano 1992, 214; he refers to Vygotsky 1986 in the last quoted sentence)

Crapanzano distinguishes between two types of shadow dialogue: one which takes place during the primary dialogue, and one afterwards.

Internal dialogue See INTERIOR DIALOGUE

Internally persuasive discourse See DISCOURSE

Interpellation According to the French MARXIST philosopher Louis Althusser, all IDEOLOGY '*hails or interpellates concrete individuals as concrete subjects*, by the functioning of the category of the subject' (1971, 162). Althusser (possibly as a result of the influence of Lacan) is making use of a technical term used to describe what happens when the order of the day in a governmental chamber is interrupted so as to allow a Minister to be questioned. The implication is that, like the Minister, individuals are interrupted and called to account – but in this case, by different ideologies. As ideology calls them – so the argument goes – so they recognize who they are. In other words: individuals come to 'live' a given set of ideological assumptions and beliefs, and to identify these with their own selves, by means of a process whereby they are persuaded that that which is presented *to* them actually represents their *own* inner identity or self. For Althusser, then, the SUBJECT is the *concrete individual* after interpellation, that is, after a sort of ideological 'body snatching'. However, Althusser believes that bodies are always already snatched; he adds that individuals '*are always-already subjects*'; even before being born 'an individual is always-already a subject' (1971, 164). According to Althusser, the only way for an individual to change this is for him or her, 'from within ideology', 'to outline a discourse which tries to break with ideology, in order to dare to be the beginning of a scientific (i.e. subject-less) discourse on ideology' (1971, 162).

On the theoretical level, the discussion by Etienne Balibar and Pierre Macherey in 'Literature as an Ideological Form' (1973) should be consulted. Some literary critics have applied Althusser's concept of interpellation to the way in which a reader adopts the 'subject' of a literary narrator or CHARACTER as the consciousness through which the literary WORK or events in it are experienced and assessed: Roger Webster refers to the reader's experience of Leo Tolstoy's *Anna Karenina*: 'The reader is drawn towards Levin and becomes through him the experiencing centre of the novel's organic vision: unless we resist such positioning by reading against the grain, it is hard to avoid the process' (1990, 82–3).

Not to be confused with INTERPOLATION. See also ENTHYMEME.

Interpolation To interpolate is to make insertions in something, and the term has been used in different ways by recent literary critics and theorists. The term *interpolated narration* (sometimes referred to as *intercalated narration*) is used to describe passages of NARRATIVE which come between two moments of action. Prince (1988, 44) points out that the epistolary novel provides many examples: letters are normally written in between dramatic events rather than during them, although *Shamela*, Henry Fielding's parody of Samuel Richardson's *Pamela*, suggested that Richardson's CHARACTERS wrote letters when they would not normally have been written in real life.

In an essay on Virginia Woolf entitled 'Virginia's Web', Geoffrey Hartman (1970) has argued that Woolf's subject is the activity of the mind, and he defines that activity as a work of interpolation: the mind is perpetually filling in gaps and adding explanatory information.

Not to be confused with INTERPELLATION.

Interpret[at]ive communities The notion of the interpretive community stems from the American critic Stanley Fish; British critics have sometimes adopted the term and modified it to *interpretative* communities so as to conform with more usual British English usage. Fish's view of the interpretive community is bound up with a related concept: that of the *interpretive strategy*:

> it is interpretive communities, rather than either the text or the reader, that produce meanings and are responsible for the emergence of formal features. Interpretive communities are made up of those who share interpretive strategies not for reading but for writing texts, for constituting their properties. In other words these strategies exist prior to the act of reading and therefore determine the shape of what is read rather than, as is usually assumed, the other way round.
>
> (1980, 14)

From this perspective an interpretive community is rather like a speech community – unified around adherence to a common set of rules which enable meaning transformations and MEDIATIONS. An interpretive strategy is thus comparable to the transformations made possible by the rules of grammar and syntax: its possession by a group of people means that they will all share a common relation to a TEXT or an UTTERANCE because they will bring the same transformational/interpretative procedures to bear on the text-to-be-interpreted or the utterance-to-be-understood.

This leads Fish to claim that when READERS interpret a text in either the same or in varying ways this is because

> members of the same community will necessarily agree because they will see (and by seeing, make) everything in relation to that community's assumed purposes and goals; and conversely, members of different communities will disagree because from each of their respective positions the other 'simply' cannot see what is obviously and inescapably there. This, then, is the explanation for the stability of interpretation among different readers (they belong to the same

> community). It also explains why there are disagreements and why they can be debated in a principled way: not because of a stability in texts; but because of a stability in the makeup of interpretive communities and therefore in the opposing positions they make possible. (1980, 15)

It will be seen that there is a potentially dangerous circularity here. Any group of individuals which reaches agreement concerning ground-rules necessary for the discussion of a text belong by definition to the same interpretive community, while any group which disagrees about these ground rules consists of members from different interpretive communities. This becomes more of a problem when we remember that two readers may reach such agreement with regard to one text while failing to do so with regard to another. Moreover, it is hard to see what might disprove the theory: if two readers share a common interpretation then they belong to the same interpretive community; if they disagree but can talk about their disagreement then they still belong to the same community; but if they cannot even find a common ground on which to talk about a difference then they belong to different interpretive communities.

In a section dealing with Fish's theory of the interpretive community in his book *Textual Power* Robert Scholes indicates other problems: only those who belong to the same community can discuss interpretive disagreements – but as they belong to the same community they shouldn't have any such disagreements to discuss. Thus only those who have no disagreements can settle them in a principled way! Scholes also points out that the inevitable multiplicity of interpretive communities (see my comments above) means that the interpretive community cannot be equated with Thomas Kuhn's PARADIGM or Michel Foucault's *ÉPISTÉMÈ* (1985, 154-6).

See also OPPOSITIONAL READING.

Intersubjectivity Used to suggest that SUBJECTIVITY is not specific to the individual but – because it is formed by common forces – is a collective phenomenon. In recent literary theory often associated with the view that the reader's experience of a TEXT is actively internalized through incorporation in his or her self rather than passively adopted or 'taken over'. As a result, a READER's view of a WORK becomes, in part, a view of him or herself: the work has been structured into the reader and is no longer a merely objective fact. Such a view gives the reader more of a creative rôle than he or she is accorded by a number of other theories, in which textual MEANING is received rather than received/constructed.

Intertextuality A relation between two or more TEXTS which has an effect upon the way in which the *intertext* (that is, the text within which other texts reside or echo their PRESENCE) is READ. Sometimes the term *transtextuality* is reserved for more overt relations between specific texts, or between two particular texts, while *intertextuality* is reserved to indicate a more diffuse peneration of the individual text by memories, echoes, transformations, of other texts. Gérard

Genette has also coined the terms *hypertext* and *hypotext* to refer to the intertext and the text with which the intertext has some significant relation.

Mikhail Bakhtin's insistence upon the DIALOGIC element in all UTTERANCES, and the range of different dialogues to be traced in literary works, drew more overt attention to the issue of intertextuality. A good example here is his extended discussion of 'the problem of quotation' in his essay 'From the Prehistory of Novelistic Discourse' (in Bakhtin 1981). He pays particular but not exclusive attention to such forms as parody and travesty and develops a theory of the linguistic HYBRID to cover them, pointing out in passing parallels with the use of parody and travesty in the modern novel (1981, 77).

Bakhtin seems himself to be an intertextual presence in Roland Barthes's and Julia Kristeva's development of theories of intertextuality. In her *Desire in Language* Kristeva defines the text as

> a permutation of texts, an intertextuality: in the space of a given text, several utterances, taken from other texts, intersect and neutralize one another. (1980, 36)

Commenting upon the concept of intertextuality in his Introduction to *Desire in Language*, Leon S. Roudiez claims that it has been generally misunderstood. According to him, intertextuality has nothing to do with matters of influence of one writer upon another or with the sources of a literary work;

> it does, on the other hand, involve the components of a *textual system* such as the novel, for instance. It is defined in [Kristeva's] *La Révolution du Langage Poétique* as the transposition of one or more *systems* of signs into one another, accompanied by a new articulation of the enunciative and denotative position. (Kristeva 1980, 15)

Roland Barthes seems partly in agreement with this position, at least so far as the distinction between intertextuality and influence is concerned. But his usage seems significantly more diffuse and all-embracing than Kristeva's. According to him, any text is an intertext.

> Any text is a new tissue of past citations. Bits of code, formulae, rhythmic models, fragments of social languages, etc. pass into the text and are redistributed within it, for there is always language before and around the text. Intertextuality, the condition of any text whatsoever, cannot, of course, be reduced to a problem of sources or influences; the intertext is a general field of anonymous formulae whose origin can scarcely ever be located; of unconscious or automatic quotations, given without question-marks. (1981b, 39)

John Frow has made a number of thought-provoking points about intertextuality which help to demonstrate how much more complex it is than some accounts have implied. His approach reminds us that intertextuality is not restricted to genesis, and that each new reading may involve a different set of intertextual relations.

> [A]ny particular construction of a set of intertextual relations is limited and relative – not to a reading subject but to the interpretive grid (the regime of reading) through which both the subject position and the textual relations are constituted. (1986, 155)

Most interesting of all, he argues that 'the text has not only an intertextual relationship to previous texts (in the case of the classics this is usually effaced) but also an intertextual relationship to itself as canonized text' (1986, 230–31). Intertextuality has thus to be seen in association with the whole complex issue of the reader's varied expectations as formed by ideological, generic, and other factors. It must also be seen, according to Frow's view, not as something which is just handed to the reader on a plate, as it were congealed into the text, but as something which the reader can to some extent control and change.

Intradiegetic See DIEGESIS AND MIMESIS

Introjection See PROJECTION CHARACTERS

Intrusive narrator A NARRATOR who breaks into the NARRATIVE to comment upon a CHARACTER, event or situation – or even to introduce opinions not directly related to what has been narrated. The term is often reserved for situations in which the intrusion is felt to break into an established narrative tone or illusion, although it is also used in a purely technical sense to describe 'own voice' comments from a narrator which may hardly be remarked by the READER because of their homogeneity with the rest of the narrative.

Iridescence See FLICKER

Isochrony Borrowed by recent NARRATIVE theorists from a term used to describe poetic rhythm, isochrony denotes an *unvarying* or an *equal* relationship between NARRATING time and STORY time. The two are not the same: if a story covers three hours and each hour is narrated by means of five thousand words, then the relationship between narrating time and story time is unvarying. But if a story covers three hours and each hour of the story takes approximately an hour to read, then the relationship between narrating time and story time can be said to be equal. It should be clear that whereas the former relationship can be measured with some degree of precision, the latter cannot. (Different readers read at different speeds, and one's speed of reading varies according to a number of factors – level of textual difficulty or interest, for example.) As Mieke Bal puts it, real isochrony (of the second type) cannot be determined precisely, although we may 'assume that . . . a dialogue without commentary takes as long in TF [FABULA-time] as it does in TS [story-time]' (1985, 70–71).

The opposite of isochrony is *anisochrony*: either a varying or an unequal relationship between narrating time and story time – normally the former.

See also DURATION.

Isotopy See TOPIC

Iterative See FREQUENCY

J

Jouissance To the surprise of many English-speaking people, this word can be found in the OED, although classified as obsolete and with examples cited from, among others, Carew and Spenser. Of the two main meanings given, that which is nearer to the current usage found amongst critical theorists is the second: Pleasure, delight; merriment, mirth, festivity. This sounds very innocent, and is clearly different from the current usage – loaned from the French – that involves *sexual* PLEASURE. The first meaning concerns the possession and use of something affording advantage, as in the *enjoyment* of a right, and jouissance is etymologically related to the word enjoyment.

Leon S. Roudiez dates the renewed critical interest in this term from the publication of Jacques Lacan's discussion of it in his 1972–3 seminar. In its French publication this sported a cover picture of the *Ecstasy of St. Theresa*, which suggests the prominent part played by sexual orgasm in jouissance. For Lacan, Roudiez claims, jouissance 'is sexual, spiritual, physical, conceptual at one and the same time' (Kristeva 1980, 16).

In Roland Barthes's *The Pleasure of the Text* jouissance is translated as *bliss*. Barthes claims that to judge a TEXT according to pleasure means that it is impossible to say either that it is good or that it is bad, because the text is 'too much *this*, not enough *that*' (1976, 13). This suggests that textual jouissance is an orgasmic experience in which the reader is so enrapt (or enwrapped) that that objectivity and distance necessary for judgement is impossible.

Kenosis See REVISIONISM

Kernel event See EVENT

Kernel function See FUNCTION

Kernel word or sentence In STYLISTICS, any word possessed of such stylistic emphasis as to colour the stylistic force of a textual unit, however defined.

L

Lack See ABSENCE

Lamella See HOMMELETTE

Langue and parole Perhaps the most important – and influential – distinction introduced by Ferdinand de Saussure in his *Course in General Linguistics*. It is now common to use the French words to represent these paired concepts in English, but one can find attempts to render them in English, with langue represented by *language* (sometimes *a* or *the* language, as in the English translation of Roland Barthes's *Elements of Semiology*, or as *language-system*), and parole as *speaking*, *speech*, *language-behaviour*, or, on occasions, phrases such as *the sum of all actual (possible) utterances*.

For the sake of clarity, where other terms are used in the following quotations, I will replace them with either [langue] or [parole] – in square brackets so that the substitution is MARKED.

According to Saussure

> If we could embrace the sum of word-images stored in the minds of all individuals, we could identify the social bond that constitutes [langue]. It is a storehouse filled by members of a given community through their active use of [parole], a grammatical system that has a potential existence in each brain, or, more specifically, in the brains of a group of individuals. For [langue] is not complete in any speaker; it exists perfectly only within a collectivity.
>
> (1974, 13–14)

A number of points need to be stressed here. First, that as the reference to 'a given community' makes clear, Saussure's mention of the minds of 'all individuals' should be taken to refer to all individuals within a particular language community. Second, that langue is supra-individual: were Martians to kidnap a single English speaker they could not extract the langue of English from him or her alone. Third, that langue is a *system*, and one that has generative power ('potential existence'). Thus, if those same Martians were able to gather every example of English speech and feed them into a super computer they still could not end up with our langue, for langue is that set of rules, that system, that is able not just to generate all those acts of English speech, but also all the *potential but as yet unuttered acts of speech* that *could* be generated by it. (However, Saussure argues [1974, 15] that although we no longer speak dead languages, we can gain access to their linguistic systems [langues]; this suggests that langue *can* be said to be accessible even to those without the ability to *generate* paroles, so long as they can understand all previously generated paroles.) Fourth, it is also the system that allows native speakers to *understand* all the acts of speech correctly generated (in other individuals) by itself.

Saussure stresses that langue is not a function of the individual speaker: it is passively assimilated by the individual and does not require premeditation (contrast speaking in a *foreign* language). Parole, on the other hand, he insists is 'an individual act' which is wilful and intellectual (1974, 14).

During the early years of the current revival of interest in the work of Saussure both Roland Barthes (1967a) and Jonathan Culler (1975) pointed out that assigning different elements to either langue or to parole was not unproblematic, but the over-arching distinction between rule and behaviour has proved extremely fertile in a range of different contexts. It may well be that the more recent growth of interest in PRAGMATICS marks the end of a rather uncritical use of the distinction: in our post-pragmatics age there are rather fewer who are prepared unreservedly to accept Saussure's contention that the 'science of language is possible only if the other elements [of parole] are excluded' (1974, 15). Culler assumes the possibility of a relatively unproblematic relating of langue and parole to Chomsky's COMPETENCE AND PERFORMANCE (1975, 9), and he seems to use these terms (here and elsewhere) relatively interchangeably.

In literary criticism this distinction has been most influential as a model. The use of the LINGUISTIC PARADIGM by STRUCTURALIST critics has led to a succession of attempts to find parallels to langue and parole in the 'system of literature'. In an essay entitled 'Structuralism and Literary Criticism' (first published in French in 1964), Gérard Genette suggested that

> literary 'production' is a *parole*, in the Saussurian sense, a series of partially autonomous and unpredictable individual acts; but the 'consumption' of this literature by society is a *langue*, that is to say, a whole the parts of which, whatever their number and nature, tend to be ordered into a coherent system. (1982, 18–19)

One of the more interesting attempts to develop this analogy is to be found in a number of successive attempts by Jonathan Culler to distinguish between a general LITERARINESS and the specific acts of literary READING this enables. Culler does not always refer directly to Saussure but sometimes to the distinction between competence and performance. However, given the comments made above this should not be too significant.

Culler argues that just as Linguistics has changed its focus with the realization that description of a finite set of sentences is no longer enough, and that 'linguistics must instead describe the ability of native speakers, what they know when they know a language', so too the study of literature must 'become a poetics, a study of the conditions of meaning' (1980, 49), and abandon the attempt merely to 'analyse a corpus of works' (1980, 50). This argument is linked to a view of literature as institution:

> Just as sequences of sound have meaning only in relation to the grammar of a language, so literary works may be quite baffling to those with no knowledge of the special conventions of literary discourse, no knowledge of literature as an institution. (1980, 49)

According to Culler, 'the conventions which make literature possible, are the same whether one adopts the reader's or the writer's point of view', and he suggests that 'as a reader oneself, one can perform all the experiments one needs' (1980, 50, 51).

Here some caveats become necessary. We seem to have moved a long way from Saussure's supra-personal system (langue) to a competence to which anyone can obtain access by means of a process of introspection that would hardly have been challenged by the New Critics. Moreover, a clear difference between linguistic and literary competence would appear to be that whereas the former normally involves *both* the generation *and* the understanding of all grammatical sentences within one's native language, literary competence for most people is limited to understanding rather than production of literary works. Furthermore, whereas langue (as Saussure pointed out – see above) is passively assimilated by the individual and does not require premeditation, literary 'competence' seems on the evidence to require an educational system to raise it to a certain level. Certainly the lowest common denominator of literary competence would appear to be a good deal lower than that of linguistic competence in our CULTURE.

Genette and Culler are by no means the only theorists to attempt to apply the langue/parole distinction to literature. In his *Introduction to Poetics* Tzvetan Todorov clearly defines poetics as the study of the literary equivalent of langue.

> It is not the literary work itself that is the object of poetics: what poetics questions are the properties of that particular discourse that is literary discourse. Each work is therefore regarded only as the manifestation of an abstract and general structure, of which it is but one of the possible realizations. Whereby this science is no longer concerned with actual literature, but with a possible literature in other words, with that abstract property that constitutes the singularity of the literary phenomenon: *literariness*. (1981, 6–7)

It is worth noting that whereas Culler sees the literary equivalent of parole to be the *analysis* of individual texts, Todorov sees these texts (he actually refers to WORKS rather than texts) *themselves* as if they were acts of literary parole – while Genette has suggested that it is the *production* of literature that constitutes the literary parole (see above). The difference – and the fact that there can be such a difference – suggests that the literary assimilation of the langue/parole distinction may be a little less straightforward than has sometimes been suggested.

Legitimation A process whereby a state, government, individual, action, or whatever is *legitimated*, that is to say, made to appear to conform with the rules or principles of the existing or dominant order.

The term is often used in recent literary criticism to suggest that one of the rôles of literary works is to present the ruling order or its values in such a light as to legitimize it. Thus Graham Holderness suggests that E. M. W. Tillyard's study *Shakespeare's History Plays* (1944)

> reproduces [Shakespeare's] plays as parables of order, or what contemporary criticism would prefer to call 'strategies of legitimation', cultural forms by means of which the dominant ideology of the Tudor state validated its own moral and political power, through the intervention of a loyal and talented subject, Shakespeare. (1992, 21–2)

Leitmotif See THEME AND THEMATICS

Lexia / lexie See FUNCTION

Linguistic paradigm Ferdinand de Saussure established a number of extremely influential analytical distinctions in his work which, while originally applied by him to the study of language, have subsequently been applied to other things. Concepts such as SYNTAGMATIC AND PARADIGMATIC relations, LANGUE AND PAROLE, and SIGNIFIER AND SIGNIFIED have all been pressed into service by theorists concerned with a range of non-linguistic phenomena.

Jonathan Culler gives an extended account of two such uses of the linguistic paradigm in his *Structuralist Poetics*: Claude Lévi-Strauss's structural analysis of MYTHOLOGY and Roland Barthes's of fashion (Culler 1975, 32–54). But once one has grasped the essential idea, the method has a potentially unending set of applications. Thus for example one can treat a meal with several courses like a sentence composed of several words: in a given culture there are all sorts of different dishes that can be chosen as the first course, but once chosen these constrain what is chosen as the second course – and so on.

Within the field of literary criticism one can refer to the way in which terminology and distinctions taken from the grammar of verbs have been applied to the study of NARRATIVE by Gérard Genette. Hence he uses *tense* to designate temporal relations between narrative and story; MOOD to designate forms and degrees of narrative 'representation'; and *voice* to designate the narrative situation or its instance (1980, 30–31). (See the entry for PERSPECTIVE AND VOICE.)

Jacques Lacan's claim that the UNCONSCIOUS is structured like a language represents a more global use of language as explanatory paradigm; other theorists have taken Noam Chomsky's distinction between COMPETENCE AND PERFORMANCE and applied it to literature: readers of literature are said to possess (perhaps to different degrees) a fundamental competence *vis-à-vis* the

READING of literary WORKS, a competence which results in certain specific performances – that is, specific readings of literary works.

In addition to applying particular distinctions originating in the study of language to non-linguistic sign-systems, those influenced by Saussure and other linguisticians have tended to place great significance on the fact of BINARY opposition itself: a fondness for binary distinctions is one of the MARKS of those using, or influenced by, the linguistic paradigm.

See also *BRICOLEUR*; HOMOLOGY; SEMIOTIC.

Lisible See READERLY AND WRITERLY TEXTS

Literariness In his essay 'The Theory of the "Formal Method"' Boris Eichenbaum quotes tellingly from Roman Jakobson's 'Recent Russian Poetry, Sketch 1', first published in Prague in 1921. In the extract quoted, Jakobson makes the polemical claim that

> The object of study in literary science is not literature but literariness – that is, that which makes a given work a work of literature. Until now literary historians have preferred to act like the policeman who, intending to arrest a certain person, would, at any opportunity, seize any and all persons who chanced into the apartment, as well as those who passed along the street. The literary historians used everything – anthropology, psychology, politics, philosophy. Instead of a science of literature, they created a conglomeration of home-spun disciplines. (Eichenbaum 1965, 107; for an alternative translation see Èjxenbaum 1971a, 8)

Eichenbaum comes back to this point in his essay 'Literary Environment':

> Literary-historical fact is a complex construct in which the fundamental role belongs to *literariness* – an element of such specificity that its study can be productive only in immanent-evolutionary terms. (Èjxenbaum 1971b, 62)

It is arguable that it is this insistence upon the specificity of literature and, thus, of the *study* of literature, which is one of the key reasons why RUSSIAN FORMALISM is formalist.

The insistence served to distinguish the study of literature not just from other disciplines such as those listed by Jakobson, but also from the study of other art-forms. This insistence upon the *specificity* and distinctiveness of literature, and of the need for it to be studied in a specific and distinctive manner, not occasionally tipped over into an insistence upon the *autonomy* of literature and of the irrelevance to Literary Studies of all references to, as Jakobson put it, environment, psychology, politics, and philosophy. It should be stressed that such developments were criticized by leading Russian Formalists. Thus Jakobson himself stated that neither he, nor Tynyanov, Shklovsky or Mukařovský ever declared that art was a closed sphere, that they had emphasized not the separation of art but 'the autonomy of the aesthetic function' (Tynyanov *et al.* 1977, 19).

See also CODES OF READING; LANGUE AND PAROLE.

Locutionary act See SPEECH ACT THEORY

Logic of the same A term which has entered into FEMINIST discourse in English from the writings of the French writer Luce Irigaray. It describes a process of argumentation whereby x is treated as equivalent to y and thus effectively subsumed into the value-scheme of y. The foremost example of this is the PATRIARCHAL assumption that the male represents a standard, and that the female is thus necessarily less than this standard, or formed in response to it. In her *Speculum of the Other Woman* (publication in English 1985), she argues that woman is forced into a subjectless position by this patriarchal logic of the same: Freud models his account of the little girl's development on his account of the little boy's development, thus presenting female sexuality as 'the negative response to the male's desire' (Millard 1989, 159).

More widely, the logic of the same is typically seen in the sort of argument that runs 'x is like y, let us consider x as if it were y, x is y'. Discussions of literary influence are particular prone to this variant form of SLIPPAGE.

Logocentrism A coinage of Jacques Derrida's, (and sometimes used interchangeably with *phonocentrism*), logocentrism refers to systems of thought or habits of mind which are reliant upon what Derrida, following Heidegger, terms the metaphysics of PRESENCE – that is, a belief in an extra-systemic validating presence or CENTRE which underwrites and fixes linguistic meaning but is itself beyond scrutiny or challenge. For Derrida, such a position is fundamentally idealist, and he argues that the dismantling of logocentrism is simultaneously the deconstruction of idealism or spiritualism 'in all their variants' (1981b, 51).

In 'Writing Before the Letter' (with which *Of Grammatology* opens), Derrida says of the *history of metaphysics* that it has

> always assigned the origin of truth in general to the logos: the history of truth, of the truth of truth, has always been . . . the debasement of writing, and its repression outside 'full' speech. (1976, 3)

Logocentrism, then, is associated by Derrida with the making of ÉCRITURE subject to speech (the English translation of écriture as writing may be misleading: see the entry for écriture).

Derrida's much-quoted assertion that '*there is nothing outside of the text*' (1976, 158), which can also be rendered in English as 'there is no outside-text', has to be read in the light of his attack on logocentrism: the TEXT cannot be assigned a meaning that is underwritten by an origin, a PRESENCE, which resides in self-validating isolation beyond the confines of the text.

See also DIFFÉRANCE.

Logos Richard Harland provides the following useful gloss on Jacques Derrida's use of this term:

> a Greek word that illuminatingly brings together in a single concept the inward rational principle of verbal texts, the inward rational principle of human beings, and the inward rational principle of the natural universe. Even more illuminating, 'logos' combines all these meanings with a further meaning: 'the Law'. For 'logos' as an inward rational principle serves to control and take charge of outward material things. (1987, 146)

It should be added that in Derrida's view the sense of security provided by a belief in logos is illusory: there are, from his perspective, no such inward rational principles.

See also PRESENCE.

Ludism From a Latin root meaning to play, *ludism* and *ludic* are used interchangeably in English with *play* and *playful* in DECONSTRUCTIONIST writing or by writers influenced by deconstructionist ideas.

The central idea behind all these usages is that once the illusion of PRESENCE has been dispensed with, READING and interpretation no longer involve a decoding that is subject to the firm discipline of some CENTRE of authority that has access to the CODE book; instead the READER can observe and participate in the free play of SIGNIFIERS endlessly generating a succession of MEANINGS none of which can claim superiority or authority. The main senses of 'play' involved here are: play as in 'to play a game', and play as in 'to play a fish' or 'to play a hose'. Vicki Mistacco expresses it as follows:

> 'Ludism' may be simply defined as the open play of signification, as the free and productive interaction of forms, of signifiers and signifieds, without regard for an original or an ultimate meaning. In literature, ludism signifies textual play; the text is viewed as a game affording both author and reader the possibility of producing endless meanings and relationships. (1980, 375)

Recent discussions of POSTMODERNISM have suggested that one important way in which it can be distinguished from MODERNISM is by reference to its more playful and unserious tone. Instead of a view of the loss of CENTRE in the contemporary world as tragic, the postmodernist views this as a justification of playfulness – exploring the potentialities of SIGNIFICATIONS without an irritable searching after final truths or unified meanings.

It should be noted that in such uses the fact that games often have *rules*, and that playing among both animals and humans is often the way in which the player is prepared for life in the non-ludic world, is not generally accorded much importance.

M

Manner (maxims of) See SPEECH ACT THEORY

Marginality Literary criticism from the earlier part of the present century focussed a certain amount of attention on to the way in which AUTHORS occupying marginal or ambiguous positions *vis-à-vis* social or national identity were often able to see beyond the accepted or CONVENTIONAL attitudes and beliefs of their time, as their marginality made it difficult for them to be – or feel – fully incorporated in any dominant system of values. MODERNIST literature in particular is characterized by its relation not only to authors who occupied various forms of marginal position, but also by its overt concern with marginality as representative of something central to modern existence. Thus we can point to the large number of major early twentieth-century authors who were (to quote the title of a book by Terry Eagleton) exiles or emigrés – caught on the margins between different CULTURES.

More recently, FEMINIST writers have focussed attention on to the way in which PATRIARCHY marginalizes female experience and thus makes male experience the determining and dominating norm. Some feminist writers have seen this process as two-edged: on the one hand, it serves to invalidate female experience and to consolidate patriarchal power through the social, cultural and political disenfranchisement of women, but on the other hand, it opens women's eyes to aspects of the functioning of patriarchy, as marginalization typically confers insight. (The person who does not belong is the person who perceives most clearly what it is to which he or she does not belong.) It is, for example, noteworthy that the most important British writer of the early twentieth century who is not an exile or an emigré in the conventional sense is Virginia Woolf; her experience of marginality involves GENDER rather than culture.

Subaltern is sometimes used is a sense very similar to marginal or marginalize: it implies that individuals may be recruited to serve in subordinate positions under a determining and defining established authority.

Jacques Derrida's concept of *supplementarity* is sometimes associated with marginality. The logic behind such an association appears to be that if all representation and interpretation requires a supplementary element – can never feed merely upon that which is to be represented or interpreted – then attention is necessarily drawn to the margins of that which is to be interpreted or represented, to the borderline between the thing itself and that which is brought to supplement it. Derrida makes the point with regard to writing:

> Writing is dangerous from the moment that representation there claims to be presence and the sign of the thing itself. And there is a fatal necessity, inscribed in the very functioning of the sign, that the substitute make one forget the

> vicariousness of its own function and make itself pass for the plentitude of a speech whose deficiency and infirmity it nevertheless only *supplements*.
>
> . . .
>
> 'The sign is always the supplement of the thing itself'. (1976, 144, 145)

See also PRESENCE.

Markedness In Linguistics a marked signifier is one which is qualified or added to in such a way as to focus or adapt its meaning. Thus *bananas* is a marked form of *banana*.

The term has been adapted in discussion of IDEOLOGICALLY charged terms. Thus to use the term 'female poet' – a marked form of 'poet' – reveals that one's concept of 'poet' involves maleness (no-one ever refers to Shelley, the well-known male poet).

Marvellous See FANTASTIC

Marxist literary theory and criticism Marxism is a MATERIALIST philosophy, one which insists upon the primacy of material living conditions rather than ideas or beliefs in the life of human beings. It sees history as, in Marx's words, 'the history of class struggle' – the history of struggle for control of the material conditions upon which life rests. It is on the basis of these material conditions, and in response to the struggle for them, that ideas, philosophies, mental pictures of the world, develop – as secondary phenomena. These secondary phenomena may provide human beings with an accurate picture of reality, including themselves and their situation, but they may not. IDEOLOGIES are all related to class positions and thus, in turn, to material conditions and the struggle for their control, but this is not to say that they provide a reliable picture of these. Traditional Marxists have laid great stress upon the distinction between BASE (or basis) and SUPERSTRUCTURE, seeing the social base as essentially economic in nature, and the superstructure as constituting the world of mental activities – ideas, beliefs, philosophies, and (in the opinion of some but not all Marxists) art and literature.

For Marxists, all is in movement, and – because there is no separate or pure realm of ideas, or values, or spiritual phenomena – all is interconnected, however complex and MEDIATED the interconnections turn out to be. The complexity of these interconnections takes, according to Marx, a characteristic form: a DIALECTICAL rather than a mechanical and purely hierarchical one. And this opens up the possibility for human beings to gain at least partial control over their life-circumstances: Marxism has traditionally been an active and interventionist philosophy not a spectatorial or passive one, although this may change with a growing suspicion of the dangers of too partisan an attitude to theory.

Marxist ideas about literature have a long history. Marx himself was extremely well-read in classical and contemporary literature, and literary allusions and references abound in his work. A number of early Marxists sought to apply Marx's ideas to literature: both in terms of the interpretation and evaluation of existing literary works, and also in terms of advice to writers and those with (or seeking) political power about what sort of literature should be encouraged. The active and interventionist nature of Marxism has recurrently led to such attempts to *use* literature for social or political ends: some of these have gained a bad press in the reviews of history – as in the case of Soviet Socialist Realism – others have received a more positive response, as in the case of Brecht's attempt to use his political theatre in the interests of social revolution. It should be noted that Marxism did not introduce the political use of art and literature to the world; there is a long tradition of such attempts – one which it is fair to say the modern academic study of literature consistently underplays and undervalues.

Marxist literary criticism has had two periods of significant influence: in the 1930s, and in the 1960s. In both periods this influence has been related to a more general interest in and commitment to Marxist ideas. Undoubtedly the most influential and important Marxist literary critic of the 1930s (and after) was the Hungarian Georg Lukács, associated in particular with a strong defence of the REALISM to which he believed his Marxism committed him, alongside a concomitant hostility on the artistic and the political level to all forms of MODERNISM. Lukács's relationship to Stalinism is complex: on the one hand, his general position *vis-à-vis* realism and modernism was in tune with the 'line' of Stalin and Socialist Realism, although this line tended to be played down as the period of the Popular Front developed. But Lukács's own position was a lot more sophisticated than that of Stalin or his henchman Zhdanov, and Lukács's very positive view of the high art of the bourgeoisie was not really equatable with Zhdanov's belief that the greatest literature in the world was then being written in the Soviet Union. In Lukács's defence it has too to be pointed out that his criticism, although generalizing in many ways, attempts to grapple with the particularities of individual works of literature in a way that was not common at this time amongst Marxist critics – although it is also true of the British critic Alick West, whose *Crisis and Criticism* was published in 1937.

Since 1960 Marxist literary criticism has reflected the diversities of Marxism in the modern world. As a generalization we can say that the less contentious it has become to see literary works in the context of their emergence and subsequent life, the more Marxist ideas have penetrated literary criticism in general. Committed modern Marxist critics are more likely than their predecessors to be engaged in the study of mediating processes: ideology, the 'political unconscious' of the American Marxist Fredric Jameson, the 'literary modes of production' of the British Marxist Terry Eagleton, and the STRUCTURES OF FEELING of the Welsh cultural theorist and novelist Raymond Williams (who is rather more difficult to situate with regard to Marxism). They are also much less likely to be happy with a straightforward relegation of literature to the realm of the superstructure. The influential French Marxist Pierre Macherey's *A Theory of Literary Production* (first published in French in 1966), for instance, by seeing the writing of literature as a form of production necessarily sees it as more than the simple reflection of economic facts that vulgar Marxism attributed to literature.

A group of Russian theorists and critics grouped around the figure of Mikhail Bakhtin has been particularly influential over the past two decades, during which time their writings from the inter-war period have reached Western European and American readers in translation for the first time. In addition to Bakhtin himself the names of P. N. Medvedev and V. N. Vološinov should be mentioned: whether Bakhtin was responsible for works published under their names is still under discussion. The extent to which these writers were Marxist is also a matter for discussion: given their situation in the Soviet Union of the 1920s and the 1930s they had to pay at least lip-service to Marxist ideas, and Bakhtin appears to have retained a Christian belief all of his life. But their writings unquestionably engage productively with Marxist ideas, seeing literature and art in its genetic socio-historical context, but paying close attention to matters of linguistic, cultural and aesthetic detail and often applying Marxist principles more rigorously than the official watchdogs of Soviet art and culture were doing. They were anti-formalists (cf. Medvedev's and Bakhtin's *The Formal Method in Literary Scholarship*, first published in 1928), but did not confuse this with a belief that formal issues were unimportant.

Literary critics or theorists describing themselves as 'Marxist-FEMINISTS', or 'STRUCTURALIST-Marxists', or seeking to combine or relate Marxism and POST-STRUCTURALISM, are now common. What we can call monolithic Marxism seems very much a thing of the past, and after the collapse of communism in eastern Europe it seems unlikely to make a particularly strong comeback.

Materialism Terry Lovell has pointed out that one of the reasons why the term materialism has suffered from 'considerable loss of definition in recent years' is that those using it have taken their lead from the natural sciences. As it has appeared that the term 'matter' has become far less precise within the natural sciences – to the extent that nowadays it is sometimes used synonymously with the word 'real' – so the consequence has been that materialism has lost that sharply defined meaning which it previously possessed.

Lovell suggests that to equate 'matter' with 'real' is unhelpful, for although it is arguable that ideas are certainly real, to define them as a *material* reality leads only to confusion. She suggests that

> Materialism is more usefully restricted to an assertion of the relationship between different levels of reality, where reality is conceived on the realist model of a multi-layered structure with different levels, or depths. (1980, 26)

Commitment to such a hierarchically layered reality in which certain levels are privileged and dominant in causal terms seems to be more fundamental to materialism than is any particular definition of what matter consists of.

Within Literary Studies a materialist position has generally indicated a belief that literary works are created and read by means, and have to be explained in terms, of pressures and influences in contrast to which the works are in some sense or other secondary rather than primary. This, it should be stressed, is a *causal* and definitely not an *evaluative* judgement. (Bricks and mortar clearly come before, and are thus primary to, houses, but this does not mean that they are more valuable than houses.)

For cultural materialism see NEW HISTORICISM AND CULTURAL MATERIALISM.

Matriarchy Government by women – either within the family or in society at large, with authority descending through the mother. Significant in recent criticism mainly as an implied rather than actualized alternative to PATRIARCHY, there have, however, been claims for the past existence of fully matriarchal CULTURES. Rather than being seen as a form of patriarchy with men replacing women, most FEMINISTS see matriarchy as embodying a very different set of alternative (and better) values.

In literature matriarchies frequently occur in feminist imaginative fictions such as utopias and science fiction.

Meaning and significance Use of the word *meaning* has been associated with literary-critical discussion for a very long time, although certain theoretical arguments have made it a rather more contentious term today than it has been in the past. The pairing of *meaning* with *significance*, however, is very much associated with one particular theorist: the American critic E. D. Hirsch. In his book *Validity in Interpretation* Hirsch defines these two terms in ways that are matched and complementary:

> *Meaning* is that which is represented by a text; it is what the author meant by his use of a particular sign sequence; it is what the signs represent. *Significance*, on the other hand, names a relationship between that meaning and a person, or a conception, or a situation, or indeed anything imaginable. (1967, 8)

The definition has to be seen in the light of increasing debate about the respective rôles and authority of AUTHOR and READER in the interpretation of literary WORKS, for it proposes a clear separation of powers: the author is responsible for meaning while significance comes from the interaction of this meaning with that which lies outside the work. Thus if Milton was telling the truth when he claimed that by his use of the 'particular sign sequence' that is 'Paradise Lost' he meant to justify the ways of God to man (interpreting this as a shorthand expression of a larger and more complex intention), then this is what the meaning of 'Paradise Lost' *is*. But if a particular reader is reminded of his or her own religious conversion when reading the poem, because an earlier reading was instrumental in effecting that conversion, then that is (part of) the work's *significance* for him or her.

A rather similar distinction is made by Michael Riffaterre, whose view of the two-stage nature of the reading process also uses these terms. According to

Riffaterre, 'before reaching the significance the reader has to hurdle the mimesis', and thus the first stage in 'decoding the poem' starts with a *heuristic reading* which is where the first interpretation takes place, 'since it is during this reading that *meaning* is apprehended'. Riffaterre's second stage is that of a *retroactive reading* during which the second interpretation takes place, and this is the truly HERMENEUTIC reading (1978, 4–5). For Riffaterre, 'units of meaning may be words or phrases, or sentences, [while] *the unit of significance is the text*' (1978, 5–6).

See also DEFERRED / POSTPONED SIGNIFICANCE.

Méconnaissance In the work of Jacques Lacan, applied to the infant's *misrecognition* of the image of itself seen in the mirror, misrecognition because, as Michèle Barrett puts it, the reflected image 'offers a *false* reflection of a whole body *Gestalt* and thus transcends the infant's own knowledge of physical dependence and psychic frustration' (1991, 103). Lacan also uses the term to refer to Freud's apparent misrecognition of 'everything that the ego neglects' (1977, 22), as well as in a more general sense.

See also BODY; MIRROR STAGE.

Mediation A stress on the highly mediated nature of human interaction and communication is common to many recent theories. Such a stress militates against simple (or simplistic) views of direct transference or copying such as are attributed by their critics to so-called *reductive* or *mechanistic* views. Central to processes of mediation are systems of *transformation*, and this means that a point at one end of a chain of complex mediations can never be accurately recaptured by an unproblematic 'reading-off' at the other.

This is clearly relevant to efforts to reach social or historical insights through the READING of literary WORKS (or to the attempt to understand literary works by means of social or historical investigations), as all sorts of mediating factors such as IDEOLOGY or literary CONVENTION have to be taken into account if these efforts and investigations are to bear fruit.

In his *The Political Unconscious* Fredric Jameson's discussion of this concept draws attention to its two-sidedness: mediation is both something the investigator *does*, by a process of *transcoding* or inventing a set of terms 'such that the same terminology can be used to analyze and articulate two quite distinct types of objects or "texts", or two very different structural levels of reality' (1981, 40), and it is also the *uncovering* of relationships independent of the work of the investigator:

> What is crucial is that, by being able to use the same language about each of these quite distinct objects or levels of an object, we can restore, at least methodologically, the lost unity of social life, and demonstrate that widely distant elements of the social totality are ultimately part of the same global historical process. (1981, 226)

In summary: the investigator both *mediates* between different levels or instances, and also *uncovers mediations* between these same levels or instances. *Both* of these elements should be borne in mind with reference to the following comment by Jameson:

> Mediation is the classical dialectical term for the establishment of relationships between, say, the formal analysis of a work of art and its social ground, or between the internal dynamics of the political state and its economic base.
> (1981, 39)

'Establishing of relationships' should perhaps be understood both as 'establishing that there are relationships' and also as 'creating relationships'.

See also HOMOLOGY; OPAQUE AND TRANSPARENT CRITICISM (mediation reading).

Mémoire involontaire See AURA

Message See FUNCTIONS OF LANGUAGE

Metacriticism 'Criticism of criticism': in other words, critical theory which has as its subject literary (or other) criticism, and which attempts to analyse and categorize examples of critical practice and to establish generally applicable principles for it. In contemporary usage the term is often replaced by *literary theory*, although this can have a rather wider meaning.

Metafiction See METALANGUAGE

Metalanguage Technically, any language used to describe or refer to another language: 'a language about a language'. One of the characteristics of human word language is that it can function as its own metalanguage; we can discuss our language in that same language (as I am doing now). This is a characteristic not shared by animal communication systems: dogs cannot bark about barking.

This has a bearing on literature and literary criticism precisely because both exploit this resource. Gérard Genette suggests a version of the LINGUISTIC PARADIGM in which the literary WORK is compared to a language and literary criticism to a metalanguage (1982, 29). However, literature can itself serve as its own metalanguage, as in *metafiction*. One consequence of this is that the play of *levels* in a literary work can be very complex, and recent study of NARRATIVE has performed an important function in making our awareness of the specificity and interrelations of different narrative levels much sharper.

A *metanarrative* can thus be either a narrative which talks about other, EMBEDDED narratives, or a narrative which refers to itself and to its own narrative procedures. Metanarratives can play an extremely important rôle *vis-à-vis* the establishment of a particular IDEOLOGICAL position in a work of fiction: Anne Cranny-Francis has pointed out, for example, how recent FEMINIST rewritings of fairy-tales can problematize the READER's relationship to familiar tales by means of metanarratives which change this relationship from a passive to an active one (1990, 89).

Metafiction is, literally, fiction about fiction. To a certain extent the term overlaps with *metanarrative* because any work of fiction which contains a metanarrative will contain a metafictional element. It is generally used to indicate fiction including any *self-referential element*, (not necessarily resulting from a metanarrative: thematic patternings can also contribute to the formation of a metafictional effect in a work). Metafiction typically involves games in which levels of narrative reality (and the reader's perception of them) are confused, or in which traditional REALIST CONVENTIONS governing the separation of MIMETIC and DIEGETIC elements are flouted and thwarted. The term is generally used with reference to relatively recent POSTMODERNIST writing, but it can have wider applications to far older work in which elements of self-observation and self-commentary can be found. The 'play within a play' in *Hamlet*, for example, inevitably introduces a metafictional element into the work, for even though there is no overt introduction of a metacommentary the audience is encouraged by Hamlet's comments on the players' performance to think about the process of dramatic illusion. The comments above on the ideological rôle of metanarrative can also be applied to metafiction, and Anne Cranny-Francis further suggests that feminist metafictions can deconstruct the GENDER ideology of popular texts (1990, 99).

See also the entry for DIEGESIS AND MIMESIS (in which some variations in Genette's definition of a number of the terms mentioned here are discussed), and the entry for FRAME.

Metalepsis According to Gérard Genette, any transgressing of NARRATIVE levels, such as when Lawrence Sterne (or his narrator) implores the READER of *Tristram Shandy* to help Mr Shandy get back to his bed (1980, 234).

Metalingual See FUNCTIONS OF LANGUAGE

Metanarrative See DIEGESIS AND MIMESIS; METALANGUAGE

Metaphor See SYNTAGMATIC AND PARADIGMATIC

Metaphysics (of presence) See PRESENCE

Metonymy See SYNTAGMATIC AND PARADIGMATIC

Microdialogue See INTERIOR DIALOGUE

Mimesis See DIEGESIS AND MIMESIS

Mirror stage In his essay 'The Mirror Stage as Formative of the Function of the I as Revealed in Psychoanalytic Experience', first delivered as an address in 1949 and reprinted in his *Écrits*, Jacques Lacan argues for this *conception* (his term) as necessary to a full understanding of 'the formation of the *I*' (1977, 1). Lacan compares the behaviour of children and chimpanzees when confronted with their own image in a mirror; even at the age at which the child is outdone by the chimpanzee in instrumental intelligence, it can recognize its own image in a mirror. But whereas this act of recognition 'exhausts itself' for the chimpanzee, the child gives forth a series of gestures in which the relation between the real and the mirrored movements is played. This, Lacan suggests, leads to the creation of an 'Ideal-I' (from Freud's *Ideal-Ich*) which 'situates the agency of the ego, before its social determination' (1977, 2).

Lacan's further discussion of the conception is complex, but what it seems to stress is that the I is thus fixed in a fictive format: our conception of ourselves is necessarily a fiction which we are then put to defend from the onset of the real. The ego should not then be regarded as centred on what Lacan calls the *perception-consciousness system*, or as organized by the 'reality principle', but on the function of *MÉCONAISSANCE* (1977, 6).

Mirror text See *MISE-EN-ABYME*

Mise-en-abyme From the French meaning, literally, to throw into the abyss. The term is adapted from heraldry, and in its adapted form generally involves the recurring internal duplication of images of an artistic whole, such that an infinite series of images disappearing into invisibility is produced – similar to what one witnesses if one looks at one's reflection between two facing mirrors. Mieke Bal recommends the term *mirror-text* for literary examples of *mise-en-abyme*, as in verbal examples it is not the whole of the WORK which is mirrored but only a part. For Bal, when the primary FABULA and the EMBEDDED fabula can be paraphrased in such a manner that both paraphrases have one or more elements in common, 'the subtext is a *sign* of the primary text' (1985, 146). The possibilities for reflexivity and self-reference opened up by such repetitions are not limited to MODERNIST art and literature, but have been utilized by artists and writers over many centuries.

For a detailed study of *mise-en-abyme* which covers both the fine arts and literature, see Lucien Dallenbach (1989).

The term is sometimes used by DECONSTRUCTIONIST writers to invoke the sense of vertigo produced by that instability of MEANING resulting from the endless play of SIGNIFIERS in a TEXT.

Misprision / misreading See REVISIONISM

Mode Apart from serving as a general synonym for 'type', mode enters into recent critical vocabulary mainly in connection with NARRATIVE theory. A usage associated with the linguistician M. A. K. Halliday equates mode with what can be termed the 'medium' of a text or 'channel of communication'. Thus a telephoned message is in a different mode from a written one.

Alternatively, alterations in narrative DISTANCE can be said to produce different modes; thus a sudden shift into irony on the part of a NARRATOR involves a change of mode.

In Gerald Prince's definition MOOD (see also the entry for PERSPECTIVE AND VOICE) consists of two sub-categories: *perspective* (or point of view) and *distance* (or mode) (1988, 54). In other words: one determines the mood of a narrative by finding out (i) what perspective on CHARACTERS and events the narrative has and (ii) how close to or distanced from these characters and events the narrative is.

It is fair to say that none of these usages is sufficiently well-established to be dominant or authoritative.

Modernism and Postmodernism Both of these terms reach beyond national-cultural and generic boundaries; they describe artistic and cultural artifacts and attitudes of (mainly) the present century which possess certain family resemblances. The term *postmodernism* can, further, be used to refer not just to art and CULTURE but also more comprehensively to aspects of modern society.

It is not easy to define modernism and postmodernism independently because the boundaries between the two terms vary according to different usages – as Andreas Huyssen, for example, points out:

> the amorphous and politically volatile nature of postmodernism makes the phenomenon itself remarkably elusive, and the definition of its boundaries exceedingly difficult, if not per se impossible. Furthermore, one critic's postmodernism is another critic's modernism (or variant thereof), while certain vigorously new forms of contemporary culture (such as the emergence into a broader public's view of distinct minority cultures and of a wide variety of feminist work in literature and the arts) have so far rarely been discussed *as* post-modern . . . (1988, 58–9)

Indeed, as Ihab Hassan points out, this indeterminacy can draw in other terms such as *avant-garde*:

> Like other categorical terms – say poststructuralism, or modernism, or romanticism for that matter – postmodernism suffers from a certain *semantic* instability. That is, no clear consensus about its meaning exists among scholars. . . . Thus some critics mean by postmodernism what others call avant-gardism or even neo-avant-gardism, while still others would call the same phenomenon simply modernism. (1985, 121)

In like vein, David Harvey has suggested that there is more continuity than difference in the movement from modernism to postmodernism, and that the latter represents a crisis within the former in which fragmentation and ephemerality are confirmed while the possibility of the eternal and the immutable is treated with far greater scepticism (1989, 116). Similarly, Alex Callinicos (1989) has argued that there is no sharp distinction between modernism and postmodernism, and that the belief that there is can be explained by reference to the particular political and cultural disappointments of the generation of 1968 in Western Europe and the United States.

Of the terms so far mentioned, perhaps the least problematic is *avant-garde*. The term comes from military terminology, and refers to the (normally small) advance guard which prepares the way for a larger, following army – what later became known as shock-troops. In the context of cultural politics the term was used in the early part of the present century to refer to movements which had the aim of assaulting CONVENTIONAL standards and attitudes – particularly but not exclusively in the field of culture and the arts. Thus Cubism, Futurism, Dadaism, Surrealism and Constructivism are all conventionally described as avant-gardist in essence.

The term *modernism* has achieved a more stable meaning during the past decade, although the attempt to, as it were, 'backdate' postmodernism to apply to works of art previously described as modernist has shaken this stability somewhat. (Thus works such as Virginia Woolf's *The Waves* and James Joyce's *Finnegans Wake* have been confidently described as *modernist* for some years, but during the last decade or so the term *postmodernist* has begun to be applied to them by some critics.) In general usage, though, modernism describes that art (not just literature) which sought to break with what had become the dominant and dominating conventions of nineteenth-century art and culture. The most important of these conventions is probably that of REALISM: the modernist artist no longer saw the highest test of his or her art as that of verisimilitude. This does not mean that all modernist art gave up the attempt to understand or represent the extra-literary world, but that it rejected those nineteenth-century standards of realism which had hardened into unquestioned conventions. Instead, the modernist art-work is possessed, typically, of a *self-reflexive* element: we may lose ourselves in the fictional 'world' of, say, *Pride and Prejudice* when reading Jane Austen's novel, but when reading James Joyce's *Ulysses* or Virginia Woolf's *The Waves* we are made conscious that we are reading a novel. (Just as we might look at a painting by Turner and lose ourselves in the scene depicted, while a painting by Picasso thrusts its 'paintingness' upon our attention.) John Frow has, however, asserted that while the modernist aesthetic is characterized by attention to the status of the utterances it produces, this does not usually extend to 'a political awareness of the social and institutional conditions of enunciation' (1986, 117).

We can compare the attack on perspective in the visual arts and on tonality in music with the attempt of various modernist writers to escape the constraints of traditional views of CHARACTER and PLOT. Thus modernism announces itself as a break with the past similar, in some ways, to the assault on traditional values associated with romanticism. One of the qualities which distinguishes modernism from romanticism, however, is a generally more pessimistic, even tragic view of the world. Generalization is dangerous here, and it seems that

British literary modernism (which, revealingly, was contributed to by many non-Britons) is perhaps more pessimistic than Continental modernism. But the work of T. S. Eliot, Ezra Pound, D. H. Lawrence, Franz Kafka, Knut Hamsun – to take some representative names – is typically characterized by a pessimistic view of the modern world, a world seen as fragmented and decayed, in which communication between human beings is difficult or impossible, and in which commercial and cheapening forces present an insuperable barrier to human or cultural betterment.

In general, modernists are hostile towards, or at least suspicious of, developments in contemporary science and technology. This is not universally the case: it is not true, for example of Vladimir Mayakovsky or of the Italian Futurists. But it is revealing that the latter have often been described as avant-garde rather than modernist. This suspicion of science and technology, in many cases directly attributable to revulsion from the use of technology to slaughter millions in the First World War, and often associated with a disgust at commercialism, is one of the clearest ways of distinguishing modernism from much postmodernism.

Here one needs to point out that an involvement in a cultural and artistic revolution does not necessarily imply political progressiveness: the work of Knut Hamsun, T. S. Eliot, Ezra Pound, Luigi Pirandello, D. H. Lawrence, and W. B. Yeats is central to modernism, but the social and political vision which can be extracted from it is more backward- than forward-looking, more conservative than progressive in political terms.

The development of modernism seems to be associated with a certain 'masculinization' of art: in contrast to the dominant position of women novelists in the nineteenth century, women take a long time to reattain this supremacy subsequent to the modernist revolution. Part of the explanation of this may relate to the association of modernism with social and geographical mobility, and the adoption of a bohemian life-style – much more difficult for women than for men, as Virginia Woolf pointed out (in the course of a different argument) in her *A Room of One's Own.*

David Harvey has argued that modernism took on multiple perspectivism and RELATIVISM as its epistemology for revealing 'what it still took to be the true nature of a unified, though complex, underlying reality (1989, 30). (Post-modernism, in contrast, tends to retain the relativism while abandoning the belief in the unified underlying reality, and David Harvey quotes François Lyotard's definition of the postmodern as 'incredulity towards metanarratives' [1989, 45] – a definition which is itself, paradoxically, something of a metanarrative; the resemblance between postmodernism and DECONSTRUCTION is strongly apparent at this point.) For the modernist, therefore, human beings are doomed to exist in a state of social – and even existential – fragmentation, while yearning (unlike the postmodernist) to escape from this situation. Here the influence of Freud is probably important, for Freud turned the attention of many writers inward, towards subjective experience rather than the objective world. On the one hand, this led to the development or refinement of important new techniques: Joyce's and Woolf's development of internal monologue and stream-of-consciousness, Eliot's refinement of the dramatic monologue. But it also tied in with a pessimistic belief in the unbridgeability of the gap between subjective experience and an objective world, the belief that 'It is impossible to say just what I mean!'. ALIENATION becomes close to a cliché in modernist literature, and it is typically associated with *urban* landscapes: we can hardly imagine T. S. Eliot's 'The Waste Land' or James Joyce's *Ulysses* set in the middle of the countryside.

Furthermore, this alienation also leads to – or is associated with – a problematizing of human individuality and identity. 'Who am I?' asks Virginia Woolf's Bernard, in her experimental novel *The Waves*, and his question is emblematic of a recurrent problem for modernist artists.

Ihab Hassan has traced the term *postmodernism* back to Frederico de Onis's use of the term *postmodernismo* in his *Antología de la Poesía Espanola e Hispanoamericana* (1934), but a letter from Charles Jencks to the *Times Literary Supplement*, 12 March 1993, points out that it was used by the British artist John Watkins Chapman in the 1870s, and in 1917 by Rudolf Pannwitz. In the same letter he adds that one of the great strengths of the word is that it implies that one has gone beyond the 'clearly inadequate' world-view of modernism, but without specifying where we are going. He also notes that as the term 'modernism' was apparently coined in the Third Century then perhaps 'postmodernism' was first used then too (Jencks 1993, 15).

The term *postmodernism* only enters Anglo-American critical discourse in the 1950s, and only in a significant way in the 1960s. At first it seems to indicate a new periodization: postmodern art or culture is that art or culture which, in the years after the Second World War, extends or even breaks with modernist techniques and conventions without reverting to realist or pre-modernist positions. But before long critics start to use the term to refer to particular cultural, artistic – or even social – characteristics irrespective of when they manifested themselves. The use of the word 'social' is significant: *post-modernism* is typically used in a rather wider sense than is *modernism*, referring to a general human condition, or society at large, as much as to art or culture (a usage which was encouraged by Jean-François Lyotard's book *The Post-modern Condition: A Report on Knowledge* [English translation, 1984]). *Post-modernism*, then, can be used today in a number of different ways: (i) to refer to the non-realist and non-traditional literature and art of the post-Second World War period; (ii) to refer to literature and art which takes certain modernist characteristics to an extreme stage; and (iii) to refer to aspects of a more general human condition in the 'late capitalist' world of the post-1950s which have an all-embracing effect on life, culture, IDEOLOGY and art, as well as (in some usages) to a generally more welcoming attitude towards these aspects.

Thus those modernist characteristics which may produce postmodernism when taken to their most extreme forms, would include the rejection of representation in favour of self-reference – especially of a 'playful' and non-serious, non-constructive sort; the willing, even relieved, rejection of artistic

AURA and of the sense of the work of art as organic whole (although David Harvey has argued that modernist [unlike postmodernist] art is essentially auratic [1989, 22]); the substitution of confrontation and teasing of the reader for collaboration with him or her; the rejection of 'character' and 'plot' as meaningful or artistically defensible concepts or conventions; even the rejection of MEANING itself as a hopeless delusion, a general belief that it is not worth trying to understand the world – or to believe that there is such as thing as 'the world' to be understood. Postmodernism takes the subjective idealism of modernism to the point of solipsism, but rejects the tragic and pessimistic elements in modernism in the conclusion that if one cannot prevent Rome burning then one might as well enjoy the fiddling that is left open to one. This and other broad definitions of postmodernism allow for the possibility of dubbing many literary and artistic works of the early part of the present century, or even of previous centuries, as to a greater or lesser extent postmodernist: the fiction of Franz Kafka, Knut Hamsun's *Hunger*, Ezra Pound's 'Cantos', and even Laurence Sterne's *Tristram Shandy*. They also open the way to seeing postmodernist elements in the work of various POST-STRUCTURALIST and deconstructive critics such as Jacques Derrida, Michel Foucault, and Jacques Lacan.

Postmodernism is characterized in many accounts by a more welcoming, celebrative attitude towards the modern world. That this world is one of increasing fragmentation, of the dominance of commercial pressures, and of human powerlessness in the face of a blind technology, is not a point of dispute with modernism. But whereas the major modernists reacted with horror or despair to their perception of these facts, in one view of the issue it is typical of postmodernism to react in a far more accepting manner. David Harvey argues that postmodernism is mimetic of social, economic, and political practices in the societies in which it appears, and he compares the superimposition of different but uncommunicating worlds in many a postmodern novel with 'the increasing ghettoization, disempowerment, and isolation of poverty and minority populations in the inner cities of both Britain and the United States' (1989, 113). Harvey also makes some very suggestive comparisons between the new organizational structures of post-Fordist capitalism (productive dispersal alongside capital concentration) and postmodernist ideology.

In our third possible usage of the term there is also a perception that the world has changed since the early years of this century. In the developed ('late capitalist') countries the advances of the communications and electronics industries have (it is argued) revolutionized human society. Instead of reacting to these changes in what is characterized as a Luddite manner, the postmodernist may instead counsel celebration of the present: celebration of the loss of artistic aura that follows what Benjamin (one of the most important prophets of postmodernism) calls 'mechanical reproduction'. In common with some much earlier avant-gardists, many postmodernists are fascinated with rather than repelled by technology, do not reject 'the POPULAR' as being beneath them, and are very much concerned with the immediate effect of their works: publication is more a strategic act than a bid for immortality. Alex Callinicos has pointed out that such accounts tend to under-represent the relations between many of the 'high modernists' and popular culture, and he instances T. S. Eliot's interest in Music Hall and Stravinsky's indebtedness to Ragtime (1989, 15).

In the field of literature it is perhaps most uncontroversial to call the following writers and their works postmodernist: John Barth, John Ashberry, Thomas Pynchon, Donald Barthelme, William Burroughs, Walter Abish, Alain Robbe-Grillet, Peter Handke, Carlos Fuentes, and Jorge Luis Borges. Two related terms describing postmodernist fiction are *fabulation* and *surfiction*. Both terms imply an aggressive and playful luxuriation in the non-representational, in which the writer takes delight in the artifice of writing rather than in using writing to describe or make contact with an extra-fictional reality.

Moment During the 1920s and 1930s the word *moment* was given a particular meaning in Virginia Woolf's writing, one similar to James Joyce's *epiphany*. A Woolfian moment typically involved the sensation of time pausing, of a number of human and/or non-human factors coming together in a unity which was both unique but which also had the power to 'speak' an inner truth or set of inner truths, both of the participants and components and to the observer. This observer of the Woolfian moment will thus simultaneously discover something fundamental about that which is observed and, at the same time, about him or herself, something profound, complete, and of visionary intensity. Woolf admired Thomas De Quincey's ability to isolate such moments in his writing: in her 'De Quincey's Autobiography' she notes that the earlier writer 'was capable of being transfixed by the mysterious solemnity of certain emotions; of realizing how one moment may transcend in value fifty years' (Woolf 1967, 6–7).

During the past two decades a rather different meaning of *moment* has emerged, one which shares an emphasis upon the self-speaking intensity that may suddenly flash out from the coming-together of a number of separate elements, but which focusses upon social-cultural and historical as well as individual and subjective forces. The 'moment' of *Left Review*, then, is that particular concatenation of forces – social, political, cultural – which provides both the impetus to found the journal and also the audience and context to appreciate and form it.

Used thus the term shares much with the rather more technical *conjuncture*, which has a somewhat more mechanical sense of social and political forces coming together like railway-lines to a junction. *Conjuncture* seems to have risen and fallen in concord with the changing authority of Louis Althusser, and although it is still encountered, the mechanistic connotations of its rigorously Latinate etymology have something of a 1970s flavour to them. (This was when the human SUBJECT had to be excluded from the heady realms of theory.)

Compare GEST; PROBLEMATIC.

Monoglossia / monologic See DIALOGIC

Monovalent discourse See REGISTER

Montage The combination of elements taken from different sources so as to produce a new whole, but (normally) one in which the original parts are clearly discernible. The OED relates the term to the cinema, but as Dawn Ades points out in her book *Photomontage*, 'manipulation of the photograph is as old as the photograph itself, which makes photomontage (a term invented by Dadaists in Berlin after the First World War) older than cinematic montage' (Ades, 1976, 7). Literary montage is associated in particular with MODERNISM, although if one wishes to use the term loosely then it can, for example, be applied to a work such as Laurence Sterne's *Tristram Shandy*, which includes various 'found texts' in its pages – including a page resembling a marbled book-cover.

Montage occupied a central rôle in debates about REALISM and modernism in the 1930s, especially in what has become known as the 'Brecht–Lukács debate'. As this debate was only published in full after the Second World War, and in English translation only in 1977, its effect was felt some time after it took place. The Hungarian Marxist Georg Lukács included montage among the sins of Expressionism and modernism in his defences of realism during this period. For Lukács montage represented the pinnacle of the symbolist movement and was correctly to be set firmly in the centre of modernist literature and thought (Bloch *et al.* 1977, 43). Brecht, in contrast, while admitting that a certain type of *anarchistic montage* succeeded only in reflecting the symptoms of the surface of things (Bloch *et al.* 1977, 72), testified to the importance of montage technique to his own work, and remarked drily that in writing his recent play he had learned more from the paintings of the peasant Breughel than from treatises on realism (Bloch *et al.* 1977, 70). Breughel's work, of course, is characterized by the sort of juxtapositions that resemble montage more than they do the classic nineteenth-century realism favoured by Lukács.

Lukács makes some rather grudging positive noises about photomontage, probably prompted by the success of those of the German anti-fascist John Heartfield. Heartfield's work also had considerable influence after the war, and confirmed that seal of left-wing approval on montage which Lukács had attempted to deny it.

The same year in which the documents of the Brecht–Lukács debate were published in English translation, David Lodge's *The Modes of Modern Writing* appeared. Lodge refers to Roman Jakobson's characterization of montage as METAPHORIC but himself argues that montage can be either metaphoric or METONYMIC depending upon context (1977, 84), a point he develops in the context of a discussion of the opening of Dickens's *Bleak House*. This piece of writing, he suggests, becomes a kind of metaphorical metonymy, and if filmed would have to be transposed into filmic montage (1977, 100).

Mood According to Gérard Genette, 'one can tell *more* or tell *less* what one tells, and can tell it *according to one point of view or another*; and this capacity, and the modalities of its use, are precisely what our category of *narrative mood* aims at' (1980, 161–2). The term is borrowed from grammatical mood, and its adoption by Genette is a good example of the reliance by NARRATOLOGISTS upon the LINGUISTIC PARADIGM.

See also MODE; PERSPECTIVE AND VOICE.

Motif See THEME AND THEMATICS

Motivated See ARBITRARY; FUNCTION

Muted In current usage applied particularly to non-dominant groups in a given society who are denied the right of and the means to expression – especially self-expression. FEMINISTS have drawn attention to the way in which women typically constitute a muted group, unable to express their real situation and thus experiencing it as an individual deviation from a proclaimed norm rather than as the common experience which it in fact is.

Elaine Showalter credits two essays by the Oxford anthropologist Edwin Ardener with the establishment of this term. According to him, she reports, muted groups must 'mediate their beliefs through the allowable forms of dominant structures' (1986, 261). Particularly important is Edwin Ardener's belief that women's beliefs and expression find expression through ritual and art, and can be deciphered by the ethnographer. Muting, in other words, is not silencing; it involves the partial silencing and suppression of valid expression, but not to the extent that this cannot be brought to the surface by the right investigator. Not surprisingly the concept has been used by feminist literary critics as an indication of the sort of READING that should be given to literary WORKS written by women in male-dominated societies.

See also ERASURE.

Myth Two of the most influential of contemporary thinkers – Claude Lévi-Strauss and Roland Barthes – have helped to revivify the concept of myth in recent times. Lévi-Strauss's discussion of myth in *The Savage Mind* helped to establish the idea of myth as *a kind of thought*, one, as he puts it, based on elements that are 'half-way between percepts and concepts' (1972, 18). This is very different from the traditional view of myth, conveniently defined by Robert Scholes and Robert Kellogg as 'a traditional plot which can be transmitted' (1966, 12).

This shift of emphasis from myth as a sort of PLOT to myth as a way of thinking with close resemblances to (along with some differences from) IDEOLOGY, can also be found in Roland Barthes's highly original *Mythologies*. Barthes's great achievement was to bring myths home to contemporary life, to make present-day European readers aware that myths were not just something that other people (remote African tribes, Russian peasants, the ancient Greeks) believed in and created – but were part of the stuff and fabric of everyday modern life in the West. *Mythologies* includes brief studies of such diverse topics as wrestling, soap powders, the face of Greta Garbo, steak and chips, and striptease. Barthes explained that for him the notion of myth explained a particular process whereby historically determined circumstances were presented as somehow 'natural', and that it allowed for the uncovering of 'the ideological abuse' hidden 'in the display of *what goes without saying*' (1973, 11). Myth, for Barthes, thus performs a NATURALIZING function, one which can be likened to an inversion of DEFAMILIARIZATION. *Mythologies* was a very influential book, leading directly to a British book entitled *Television Mythologies* (1984) and indirectly to a rather different usage of the word 'myth'.

Clemens Lugowski's *Form, Individuality and the Novel*, first published in German in 1932 and only recently published in English, makes use of the concept of the *mythic analogue* in a way reminiscent of, though predating, the work of Barthes and Lévi-Strauss. Introducing the modern reissue of Lugowski's work, Heinz Schaffler notes that 'Lugowski's inquiry implies that while the original vitality of myth has faded, the remnants of mythic thought have gone over into aesthetic structures', and this remnant Lugowski dubs the mythic analogue (1990, xiii). As with Barthes and Lévi-Strauss, myth for Lugowski involves a way of representing reality. By reference to the *Decameron* he suggests that the mythic analogue involves a 'view of the world as a form of timeless, static existence' (1990, 42) – which reminds us of Barthes's view of myth as a transformation of history to a sort of common sense.

N

Nachträglichkeit From the German of Sigmund Freud: delayed effect or after-experience. Central to much of Freud's work is the belief that reactions to disturbing experiences (witnessing the sexual act of one's parents, for example) are not immediate but long-term and highly MEDIATED. Both Jacques Derrida and Jacques Lacan draw on this concept (in, for instance, the former's 'Freud and the Scene of Writing' and the latter's 'The Function and Field of Speech and Language in Psychoanalysis'). An insistence upon the long-term and indirect nature of chains of cause and effect is central to much recent theory – compare the entries for MEDIATION and OVERDETERMINATION.

Naked-ape-ism See BIOLOGISM

Name-of-the-Father According to Jacques Lacan 'the attribution of procreation to the Father can only be the effect of a pure signifier, of a recognition, not of a real father, but of what religion has taught us to refer to as Name-of-the-Father' (1977, 199). He further argues that Freud linked this signifier of the Father

> as author of the Law, with death, even to the murder of the Father – thus showing that if this murder is the fruitful moment of debt through which the subject binds himself for life to the Law, the symbolic Father is, in so far as he signifies this Law, the dead Father. (1977, 199)

According to Lacan the child enters language during the Oedipal phase, and it is at this time that language as it were INTERPELLATES it – to use a somewhat anachronistic term. The Name-of-the-Father constitutes the child's authority, and comes with the acquisition of language, but it is simultaneously an authority that is dead.

More loosely, the term is used to refer to any (normally PATRIARCHAL) external and unchallenged source of authority.

Narrated monologue See FREE INDIRECT DISCOURSE

Narratee The 'target' at whom a NARRATIVE is directed. A narratee is not just the individual by whom a narrative is received; there has to be some evidence that the narrative is actually intended for a particular goal for it to count as the (or a) narratee. This leads Prince (1988) to argue that the narratee must be inscribed in the TEXT. He argues that there is a difference between a narratee and both the READER and the IMPLIED READER. In the final section of James Joyce's *Ulysses* therefore, if we follow Prince on this matter, the narratee is Molly Bloom herself, but the implied reader is rather a person who can, for example, pick up the classical analogies contained within the text of *Ulysses* as a whole, with the real reader being anyone who actually reads the novel.

Some narratologists distinguish between *intra-fictional narratees* (such as Molly Bloom, or Lockwood in *Wuthering Heights*), who are fictional personae within the works concerned as well as having a narrative addressed to them, and *extra-fictional narratees*, who are not.

Narration This is a rather slippery term in contemporary NARRATIVE theory, and is given different weight by different theorists. By some it is used as a synonym for narrative, by others as the act or process whereby a narrative is produced. The second of these is the definition chosen by Shlomith Rimmon-Kenan, for whom narration is both (i) the *communication* process in which the narrative as message is transmitted and (ii) the *verbal* nature of the medium used to transmit the message (1983, 2). For Rimmon-Kenan the communication process involved in narration is a double one, both contained in the TEXT (Marlow narrates his STORY to those listening to him in *Heart of Darkness*) as well as involving the text (Joseph Conrad is engaged in narration when writing *Heart of Darkness* for READERS to read) – although for her the former of these processes is the more important (1983, 3).

Michael J. Toolan discusses a number of alternative usages, and suggests his own: narration is 'the individual or "position" we judge to be the immediate source and authority for whatever words are used in the telling' (1988, 76).

Compare the distinctions listed in the entry for ENUNCIATION and its cognates.

Narrative Gerald Prince defines this term as 'the recounting of one or more real or fictitious events' but as 'product and process, object and act, structure and structuration' (1988, 58). So far as other theorists are concerned, however, it almost seems a case of perm any two from this list of six.

Thus Gérard Genette points out that the word *narrative* (in French, *récit*) can refer to three separate things: either the oral or written narrative statement that undertakes to tell of an EVENT or events; or the succession of real or fictitious events that are the subject of the DISCOURSE, with their varied relations; or, finally, the act of narrating (1980, 25–6). In his own usage he reserves the word narrative for the first of these three, while the second he refers to as STORY or DIEGESIS and the third as *narrating*. This makes good sense, but others have suggested alternative usages and one should be prepared for the possibility that the term may be used in any one or more of the three alternatives suggested by Genette.

On two points there is, however, agreement. First that a narrative must involve the recounting of an event or events, otherwise it is not a narrative but a description. And second that these events can be either real or fictitious. The person telling the TV News what happened in an accident in which he or she was involved, is as much delivering a narrative as is the person telling a joke, or the Marlow of Joseph Conrad's *Heart of Darkness*.

See also FUNCTION; REGISTER.

Narrative levels See STORY AND PLOT

Narrative movements See DURATION

Narrative situation Used by Mieke Bal in a technical sense, and fixed according to the answers evinced by a set of typical questions. Is the NARRATOR a CHARACTER or not? Does the narrator exist within the world of the STORY? Is the NARRATIVE FOCALIZED through the narrator? (1985, 126). In other words, narrative situation has nothing to do with whether the narrative is ostensibly delivered in Paris or Rome, or whether the narrator narrates in a sitting or standing position, but is defined according to the narrator's relationship to the narrative and the story. From this perspective the narrative situations in Joseph Conrad's *Heart of Darkness* and Woody Allen's film *Broadway Danny Rose* have certain striking similarities, even though the setting, story, and action of the two WORKS have precious little in common.

Narratology According to Mieke Bal, 'narratology studies narrative texts only in so far as they are narrative' (1985, 126). In other words, a TEXT such as *The Catcher in the Rye* can be studied in a number of ways, but not all of these belong to narratology or literary criticism. Moreover, a literary-critical study of a text is not necessarily the same as a narratological study of it. If the text is studied as a source for information about the problems of adolescence then that is not primarily a literary-critical approach to the text – although a literary-critical approach could involve a concern with its treatment of adolescent problems – nor is it a narratological approach. Narratology is concerned only with the issue of how the EVENTS which make up this particular STORY are narrated.

Gerald Prince suggests variant definitions of this term, and in addition to one similar to that advanced by Bal, he suggests that narratology can refer to the STRUCTURALIST-inspired theory which studies the functioning of NARRATIVE in a medium-independent manner, and he attempts to define both narrative COMPETENCE as well as what narratives have in common and what enables them to differ from one another (1988, 65).

Narrator Whereas Gerald Prince describes the narrator as 'the one who narrates' (1988, 65), and Katie Wales as 'a person who narrates' (1989, 316), Mieke Bal stresses that for her the narrator is the narrative agent, 'the linguistic subject, a function and not a person, which expresses itself in the language that constitutes the text' (1985, 119). The tension between these two approaches pin-points a problem: whereas the term evokes a sense of a human individual for most people, many NARRATIVES do not stem from recognizably human or personified sources, but from a SUBJECT position within the text. Monika Fludernik has argued for reserving the term *narrator* for 'those instances of subjective language that imply a *speaking* subject: the personal pronoun *I*, addresses to the narratee, metanarrative commentary and evaluation' (1993, 443), but many theorists continue to use the term, as does Bal, to indicate a narrative agent of any form, personified or not.

The issues involved become clearer when we remember that the narrator must be distinguished not just from the real author, but also from the implied author. This is most obvious in the case of a *personified narrator*: Charles Marlow, who we meet with in four of Joseph Conrad's fictions, is neither Joseph Conrad nor that authorial presence and CENTRE that we sense (according to some commentators) in each complete fiction of Conrad's, and which some prefer to name the implied author. But the same point can be made about the narrator of George Eliot's *Middlemarch*, who is neither George Eliot (or Marian Evans!) nor the 'sense of an author' we get when reading *Middlemarch.* (Although narrator and implied author are more difficult to distinguish in *Middlemarch* than they are in, say, Conrad's *Lord Jim.* See the discussion of the term *career author* in the entry for AUTHOR.) But if the narrator of *Middlemarch* is not a real or implied author, it does not follow that he/she is (as Wales has it) 'a person' – otherwise I would not be driven to the awkwardness of a formulation like 'he/she'. That unwillingness to refer to the narrator of *Middlemarch* as 'he' or 'she' – or, come to that, as 'it' – reveals that although many narrators assume certain human characteristics, they must sometimes be distinguished from human individuals. Indeed, it may be argued that (like literary CHARACTERS) they need to be distinguished from human individuals even when they are personified and appear most human – as with, for example, Charles Marlow. Moreover, even if a narrator may be personified, he or she may still be relatively anonymous. The FRAME narrator of Henry James's *The Turn of the Screw* is so anonymous that there is no firm evidence as to whether we are dealing with a male or a female.

Level of personification is thus one important element in distinguishing between narrators, but it is not the only one. Other important elements are: *narrative level* (does the narrator belong to the same 'reality' as the characters, or is he/she *extradiegetic*? (See the entry for DIEGESIS AND MIMESIS.) Does the narrator *participate in the story* fully, partially, or not at all? (The narrator may be fully personified and portrayed on a REALISTIC level, but merely recounting a STORY which he/she observed without personal involvement.) Is the narrator *perceptive or obtuse*? *Wuthering Heights* has one (relatively) perceptive narrator and one obtuse one; *Huckleberry Finn* and *Gulliver's Travels* have narrators who are at times perceptive, at times obtuse. Is the narrator *overt or covert*? In other words, are we aware of a narrating subject or does the text seem transparent, giving us a view of character and action which so occupies our attention that we are not conscious of any narrator? Is the narrator *reliable or unreliable*? Do we believe everything that the narrator tells us, or suspect that either deceit or obtuseness on his/her part requires us to see more than he or she does?

Naturalization See DEFAMILIARIZATION

Negotiation See EXCHANGE

Négritude A neologism coined by the Martinique writer Aimé Césaire in his *Cahier d'un Retour au Pays Natal* (*Notes on a Return to the Land of My Birth*) in 1939. One of the definitions of the term provided by Césaire is simple and all-embracing: '*Négritude* is the simple recognition of the fact that one is black, the acceptance of this fact and of our black destiny, our history, and our culture' (Kesteloot 1968, 80). Following Césaire, the term was applied to a movement of Black (mainly African) writers resident in Paris, and was in conscious opposition to the French colonial policy of integrating colonial peoples and their cultures into French culture. The Senegalese writer Léopold Sédar Senghor was an influential figure in the movement.

The term was taken up by some Black American writers in the 1960s, for whom it became a shorthand way of celebrating Blackness and Black CULTURE. During this time its use generally betokened an anti-rational, anti-colonialist standpoint, the associations its use by Black writers in Paris had gathered by the time that visiting Black American writers encountered it. It suggested the anti-rational in as much as its use went hand in hand with a rejection of the rational in favour of more mystical, collective and emotional forms of understanding; anti-colonialist in as much as Black Americans identified their own situation as oppressed group in the United States with that of Black Africans fighting against European domination. Much Black American writing of the 1960s and 1970s used these associations to label and LEGITIMIZE a sort of writing that broke with what were seen as culturally alien forms of expression. It had its critics, however, who argued that many negative STEREOTYPES of Black people were perpetuated by the term.

New Historicism and cultural materialism Groupings of critics and theorists who have rejected the synchronic approaches to CULTURE and literature associated with STRUCTURALISM and who have attempted to provide more adequate answers to the range of problems associated with the tensions between aesthetic, cultural, and historical approaches to the study of a range of different sorts of TEXT. Most of those known as New Historicists (some of whom have gone on record with their preference for the term 'cultural poetics') are from North America, while cultural materialism is by and large a British phenomenon. On occasions, however, New Historicism is used as an umbrella term to include members of both groupings of scholars.

The writings of Michel Foucault and Raymond Williams constitute a major influence on the New Historicists, who have succeeded in defining (or suggesting) new objects of historical study, with a particular emphasis upon the way in which causal influences are mediated through discursive practices (see the entry for DISCOURSE).

A key figure in the rise of the New Historicism is the Renaissance critic Stephen J. Greenblatt, and in his recent collection of essays *Learning to Curse* (1990) he admits that for him the term describes less a set of beliefs and more

> a trajectory that led from American literary formalism through the political and theoretical ferment of the 1970s to a fascination with what one of the best new historicist critics [Louis A. Montrose] calls 'the historicity of texts and the textuality of history'. (1990, 3)

Elsewhere he describes the New Historicism as a practice rather than a doctrine (1990, 146). Greenblatt sees the New Historicism's creation of 'an intensified willingness to read all of the textual traces of the past with the attention traditionally conferred only on literary texts' (1990, 14) to be central to its value. Thus in a study of a design by Dürer for a monument to commemorate the defeat of peasants involved in protest and rebellion, Greenblatt notes that intention, genre and historical situation all have to be taken into account, as all are social and IDEOLOGICAL and must be involved in any 'reading' of the design (1990, 112). He continues:

> The production and consumption of such works are not unitary to begin with; they always involve a multiplicity of interests, however well organized, for the crucial reason that art is social and hence presumes more than one consciousness. And in response to the art of the past, we inevitably register, whether we wish to or not, the shifts in value and interest that are produced in the struggles of social and political life. (1990, 112)

The New Historicist, in other words, has as much to say about the reading of texts as about their composition.

For those who like negative definitions, Greenblatt cites three definitions of the word 'historicism' from *The American Heritage Dictionary*, all of which he sees to be counter to the practice of New Historicists:

> 1. The belief that processes are at work in history that man can do little to alter.
> 2. The theory that the historian must avoid all value judgments in his study of past periods or former cultures.
> 3. Veneration of the past or of tradition. (cited in Greenblatt 1990, 164)

Although Greenblatt and other New Historicists pay tribute to the work of various POST-STRUCTURALISTS, the anti-formalist element in their work clearly distances them from important aspects of post-structuralism.

The term 'historicist' is sometimes used in a pejorative sense which is unconnected with New Historicism. Historicist in this sense implies the view that human, social or cultural characteristics are determined in an absolute sense by historical situation; historicism in this sense is thus a form of REDUCTIONISM as the human, the social and the cultural are collapsed back into the historical. Thus the title of an essay by Louis Althusser, 'Marxism is not a historicism', rests on such a definition of historicism (the essay is included in Althusser & Balibar 1977).

Graham Holderness has produced a useful checklist of what he considers to be the differences between the (mainly British) cultural materialism and the (mainly North American) New Historicism. According to him:

> Cultural materialism is much more concerned to engage with contemporary cultural practice, whereas New Historicism confines its focus of attention to the past; cultural materialism can be overtly, even stridently, polemical about its political implications, where New Historicism tends to efface them. Cultural materialism partly derives its theory and method from the kind of cultural criticism exemplified by Raymond Williams, and through that inheritance stretches its roots into the British tradition of Marxist cultural analysis, and thence into the wider movement for socialist education and emancipation; New Historicism has no sense of a corresponding political legacy, and takes its intellectual bearings directly from 'post-structuralist' theoretical and philosophical models. . . . Cultural materialism accepts as appropriate objects of enquiry a very wide range of 'textual' materials [. . . whereas] New Historicism concerns itself principally with a narrower definition of the 'textual': with what has been *written* . . . (1991, 157)

Clearly, then, both terms denote relatively loosely defined schools which approach literary (or other) TEXTS contextually.

However, according to Alan Sinfield, the 'rough programme' of cultural materialism involves the placing of a text in its (plural) context*s*:

> a strategy [which] repudiates the supposed transcendence of literature, seeking rather to understand it as a cultural intervention produced initially within a specific set of practices and tending to render persuasive a view of reality; and seeing it also as re-produced subsequently in other historical conditions in the service of various views of reality, through other practices, including those of modern literary study. (1992, 22)

In his book *Shakespeare Recycled* (1992), Graham Holderness suggests that the New Historicists have preferred to 'reproduce a model of historical culture in which dissent is always already suppressed, subversion always previously contained, and opposition always strategically anticipated, controlled and defeated' (1992, 34). By implication, therefore, the cultural materialists see a culture far more as a battlefield: riven with struggles, tensions, and contradictions on a number of different planes. Alan Sinfield, the title of whose important cultural-materialist book *Faultlines: Cultural Materialism and the Politics of Dissident Reading* (1992) is itself indicative of this view, agrees with Holderness in tracing this crucial difference back to the work of Raymond Williams:

> Much of the importance of Raymond Williams derives from the fact that at a time when Althusser and Foucault were being read in some quarters as establishing ideology and/or power in a necessarily unbreakable continuum, Williams argued the co-occurrence of subordinate, residual, emergent, alternative, and oppositional cultural forces alongside the dominant, in varying relations of incorporation, negotiation, and resistance. (1992, 9)

Thus while we can picture the New Historicist standing safely on the shore and gazing at the sea of history (at least this is how the cultural materialist sees the matter), the cultural materialist believes that his or her head is already under water, that a historical approach requires that the investigator take full account of his or her historical location and of the historical life of literary WORKS (including their life in the present time of the investigator), not just the historical situation of the work's creation and composition. It should be added that although a New Historicist such as Greenblatt has also acknowledged an important intellectual debt to the work of Raymond Williams, this debt has (at least according to the cultural materialists) carried with it less of a political imperative than it has for the cultural materialists.

The following scholars have been associated with New Historicism: Stephen Greenblatt, Louis Montrose, Jonathan Goldberg, Leonard Tennenhouse, Stephen Mullaney, and Hayden White. The best-known cultural materialists are Jonathan Dollimore, Alan Sinfield, Lisa Jardine, Graham Holderness, Catherine Belsey, and Francis Barker.

See also CIRCULATION; EMPLOTMENT; EXCHANGE; NEGOTIATION; RESONANCE.

Norm According to Jan Mukařovský, the basic prerequisite for an aesthetic norm is not its statability, 'but the general consensus, the spontaneous agreement, of the members of a certain community that a given esthetic procedure is desirable and not another' (1964, 44). Members of the PRAGUE SCHOOL such as Mukařovský thus use *norm* in a manner which is very close to CONVENTION; certainly Mukařovský's account of the way in which breaking norms involves DEFAMILIARIZATION, is very close to Roman Jakobson's (1971) account of the way in which too rigidified a system of conventions leads to AUTOMATIZATION of response.

The *normative fallacy*, according to Pierre Macherey, is apparent when criticism attempts to modify a literary work in order that it may be more thoroughly assimilated, 'denying its factual reality as being merely the provisional version of an unfulfilled intention'. For Macherey, the normative fallacy is a variety of a more fundamental fallacy, the *empiricist fallacy*, witnessed whenever criticism 'asks only how to *receive* a given object' (1978, 19).

Nouvelle critique The French term *nouvelle critique* is not to be confused with the Anglo-American New Criticism, although it is sometimes rendered into English as the (French) New Criticism. It was brought to the attention of a larger public through a polemical attack on Roland Barthes's book *Sur Racine* (1963) entitled *Nouvelle Critique ou Nouvelle Imposture?*, written by Raymond Picard and published in 1965. *Nouvelle critique* was a term Picard adopted from the critics he was attacking (Barthes served as the main focus of his rage), and which he attempted to turn against them. Picard represented a highly conservative form of criticism, and saw the work of critics such as Barthes, Lucien Goldmann, Georges Poulet, Jean Starobinski and Jean-Pierre Richard as a threat to the fundamental values of scholarly criticism. Picard attacked Barthes in particular for failing to enter into the author's (Racine's) intention, of relying upon illegitimate psychoanalyses, of dragging in sexuality everywhere in his discussion of Racine – and so on. (For a full discussion of Picard's attack, and of the *nouvelle critique* in general, see Doubrovsky [1973].)

Nucleus See EVENT

Object-relations theory / criticism According to Nancy Chodorow:

> Object-relations theory, rooting itself in the clinical, requires by definition attention to the historically situated engagement with others as subjects, an attention to the processes of transference and counter-transference that constitute the clinical situation and do so by bringing each person's relational history to it.

> This theory argues, through clinical example, that people are fundamentally social, even that there is a fundamental social 'need'. (1989, 148)

The theory is post-Freudian in the sense that it finds an emphasis on object-relations in Freud's work but sees the need to re-emphasize this in the face of subsequent attempts to counter or reduce it.

The objects in object-relations theory are those with which the human infant engages in the course of its first year of life, a year in which mother-child relations are of paramount importance. A key figure in the theorizing of the infant's object-world is Melanie Klein, to whose work object-relations theory is largely indebted. Klein argues that

> The development of the infant is governed by the mechanisms of introjection and projection. From the beginning the ego introjects objects 'good' and 'bad', for both of which the mother's breast is the prototype – for good objects when the child obtains it, for bad ones when it fails him. But it is because the baby projects its own aggression on to these objects that it feels them to be 'bad' and not only in that they frustrate its desires: the child conceives of them as actually dangerous – persecutors who it feels will devour it, scoop out the inside of its body, cut it to pieces, poison it – in short, compassing its destruction by all the means which sadism can devise. (1988, 262)

Object-relations theory has been of particular interest to FEMINISTS as it has offered an alternative to such posited Freudian universals as penis-envy and the castration complex; Melanie Klein's work suggests an initial delight in the female on the part of both male and female infants, and her concentration on mother-child bonds has also been of great significance to object-relations criticism, particularly in its tracing of female-female relations in literary works. Also important has been its provision of a model for the internalization and objectification of interpersonal relationships.

Some literary critics associated with object-relations theory are Elizabeth Abel, Joan Lidoff, Marianne Hirsch, and Margaret Homans.

Objet a / Objet A Otherwise *Objet petit a / Objet grand A*. Coinages of Jacques Lacan. According to Lacan's translator Alan Sheridan, Lacan insisted that these terms should not be translated and refused to comment on them, 'leaving the reader to develop an appreciation of the concepts in the course of their use'. Sheridan does provide the information that the little 'a' stands for *autre* or OTHER, a concept developed out of the Freudian 'object', and that 'the "*petit a*" (small "a") differentiates the object from (while relating it to) the "*Autre*" or "*grand Autre*" (the capitalized "Other")' (1977, xi). For Lacan,

> The *object a* is something from which the subject, in order to constitute itself, has separated itself off as organ. This serves as a symbol of the lack, that is to say, of the phallus, not as much, but in so far as it is lacking. It must, therefore, be

> an object that is, firstly, separable and, secondly, that has some relation to the lack. (1979, 103)

According to Ellie Ragland-Sullivan,

> the *objet a* is any filler of a void in being, which, because of its indispensable function in filling up that void, quickly provides a consistency, palpable in repetitions. (Ragland-Sullivan 1992a, 58)

In another context she adds that

> The *objet a* – a plus value or excess in *jouissance* – enters the fantasy as '*the object which causes the desire of a subject and limits its jouissance*'. (Ragland-Sullivan 1992b, 175; the quotation is from Brousse 1987, 116)

For JOUISSANCE, see the separate entry.

Obstination According to Monika Fludernik, Paul Ricoeur's translation of a term used by Harald Weinrich to indicate the READER'S tendency to continue to FRAME a passage of NARRATIVE in a consistent manner (e.g. as FREE INDIRECT DISCOURSE) unless prompted by TEXTUAL or contextual features to shift to a new frame.

Opalescence See FLICKER

Opaque and transparent criticism Terms coined by A. D. Nuttall in his *A New Mimesis* to distinguish what he terms two languages of criticism: 'the first "opaque", external, formalist, operating outside the mechanisms of art and taking those mechanisms as object, the second "transparent", internal, realist, operating within the "world" presented in the work' (1983, 80). Thus a statement such as 'In the opening of *King Lear* folk-tale elements proper to narrative are infiltrated by a finer-grained dramatic mode' involves the use of opaque critical language, whereas 'Cordelia cannot bear to have her love for her father made the subject of a partly mercenary game' involves the use of transparent critical language (Nuttall's own examples).

In an article about the meanings of TV documentary John Corner and Kay Richardson suggest comparable terms: *mediation reading* and *transparency reading*. A mediation reading is one in which the READER is conscious of the intention or motivation behind what is being read, whereas a transparency reading is one in which 'comments are made about the depicted world as if it had been directly perceived reality' (Corner & Richardson 1986, 149).

Open and closed texts The theorist who has done most to popularize the idea that TEXTS can be either open or closed is Umberto Eco. His formulation is, however, quite complex, and he defines 'open' and 'closed' in rather unexpected ways. As a result, the terms are often used in a sense that is the reverse of what he recommends.

The problem is very clear in the following discussion by Eco of texts such as Superman comic strips and novels by Ian Fleming (the creator of 'James Bond'). For Eco, such texts are closed precisely because they are open to any sort of READING:

> Those texts that obsessively aim at arousing a precise response on the part of more or less precise empirical readers . . . are in fact open to any possible 'aberrant' decoding. A text so immoderately 'open' to every possible interpretation will be called a *closed* one. (1981, 8)

In contrast, where the AUTHOR of the text has envisaged the rôle of the READER at the moment of generation, according to Eco the text is, paradoxically, open to successive interpretations. These interpretations and reinterpretations, according to Eco, echo one another, and operate within certain textually imposed constraints (unlike responses to what he calls closed texts).

> You cannot use the [open] text as you want, but only as the text wants you to use it. An open text, however 'open' it be, cannot afford whatever interpretation. (1981, 19)

At the root of Eco's argument here lies a relatively traditional belief in the possibility of distinguishing between correct and incorrect interpretations of a certain sort of valued text.

> [I]t is possible to distinguish between the free interpretative choices elicited by a purposeful strategy of openness and the freedom taken by a reader with a text assumed as a mere stimulus. (1981, 40)

The theory thus depends upon two sorts of distinction: between different sorts of texts, and between different sorts of response to, or interpretation of, texts. For Eco, open texts, works of art, are *in movement*, another coinage of his by which he seeks to describe their power to generate a never-ending series of new but valid reading experiences and interpretations.

Eco's discussion is thought-provoking and fruitful, but the complexities of a terminology that requires us to remember that a closed text is open to any sort of response and that an open text constrains what the reader does to it, have ensured that his suggested definitions have not obtained wider popularity – even though the terms *open* and *closed* are regularly and loosely used in discussions of literary and other texts.

See also READERLY AND WRITERLY TEXTS.

Opponent See ACT / ACTOR

Oppositional reading A reading which rejects and seeks to undermine the terms overtly or implicitly proposed by a given TEXT for acceptance by the READER prior to and during its own reading. Thus an oppositional reading of Mark Twain's *Huckleberry Finn* might consciously seek to read the closing section dealing with the boys' treatment of 'Nigger Jim' in a way that the text tries to exclude: as an attempt to conciliate precisely those racist views which the novel has earlier attacked. An oppositional reading of Henry James's *The Turn of the Screw* might view this text as reflective of STEREOTYPED views of women as hysterical, unreliable, mysterious, corrupted by repressed passion, and so on.

The term 'hermeneutics of suspicion', which is associated with Paul Ricoeur, denotes a very similar process, although it lays more stress on interpretation than on the reading process and focuses attention on the unsaid rather than the said. (Compare the entry for ABSENCE.)

See also HERMENEUTICS.

Optimal reader See READERS AND READING

Orality A term used to denote an extended complex of elements associated with oral CULTURES – that is, cultures either unaffected by literacy and the written word or only marginally affected by them. One of the most influential (and accessible) theoretical contributions to the study of orality is Walter J. Ong's *Orality and Literacy* (1982), in which Ong argues that in a predominantly oral culture thought differs from the thought typical of a culture characterized by universal or near-universal literacy. The thought and expression of an oral culture, according to Ong, are *additive rather than subordinative*, *aggregative rather than analytic*, *redundant or 'copious'*, *conservative or traditionalist*, *close to the human lifeworld*, *agonistically toned* (that is, polemical and emotive), *empathetic and participatory rather than objectively distanced*, *homeostatic*, and *situational rather than abstract* (1982, 37–57).

Ong's arguments have not gone unchallenged, and there is a growing literature concerned with debates around the concept of orality. These are of relevance to Literary Studies in a number of ways. For example: it seems clear that poetry emerged in an oral culture, and that some of its inherited characteristics owe something to the thought-patterns of an oral culture. Understanding more about an oral culture may help us to understand more about both the poetry of the distant past and also of more recent periods.

Orchestration In the work of Mikhail Bakhtin, part of an analogy between musical and novelistic structure (see also POLYPHONY). The different VOICES active in a given CULTURE at a given time can be orchestrated so as to reveal different aspects alone and in harmony/disharmony with one another.

Order Both the temporal succession of events in a STORY and also the placing in linear succession of EVENTS in a NARRATIVE.

See also ACHRONY; ANACHRONY; ANALEPSIS; PROLEPSIS.

Organic intellectuals See INTELLECTUALS

Organicism In literary criticism, particularly during the heyday of the New Critics, it came to be an item of faith that the work of literature (or art) had to be treated as an *organic* structure – that is, having the qualities of a living unit with its parts organically rather than mechanically related.

During the same time F. R. Leavis extended the analogy from the individual literary work to literature as a whole; in his essay 'Literature and Society' he argues that

> A literature . . . must be thought of as essentially something more than a separation of separate works: it has an organic form, or constitutes an organic order, in relation to which the individual writer has his significance and his being. (1962a, 184)

Leavis goes on to argue that this approach is to be distinguished from that of the MARXIST, in that it stresses not economic and material determinants but intellectual and spiritual ones, and that although it accepts that material conditions have an enormous importance, 'there is a certain measure of spiritual autonomy in human affairs' (1962a, 184). Leavis also made much of the concept of 'the organic community', most notably in *Culture and Environment* which was written with Denys Thompson and first published in 1933, although directly comparable ideas can be found in the essay 'Literature and Society' (and in many other places in Leavis's writing) as well. The conservative-nostalgic concept of the organic community in Leavis's work owes much to T. S. Eliot. It represents a lost unity in sharp contrast to the alleged disintegration and divisions of modern urban society, a unity both among human beings and between human beings and their environment, in which a CULTURE shared by all the community's members and interwoven with the realities of their daily lives could be found.

It is necessary to fill in this historical sketch of earlier usages of *organic* because it is in reaction to these that a more recent usage is based. Representative is a discussion by Christopher Hampton of the passage quoted above from Leavis's 'Literature and Society'. Hampton's objection to Leavis's position centres upon the fact that as a result of the terms he uses 'history vanishes; and with it the influence of the changing conditions of material existence which determine not only the ways in which people's lives are shaped but also the cultural products of their thinking, including literature' (1990, 50). Hampton's critique is sharper and less conciliatory than that of others – notably Raymond Williams in a number of his studies – but it sums up a particular sort of reaction against certain of Leavis's ideas. And in so doing it sums up what is meant by the current, pejorative use of the term *organicist*: viewing either literature, or art, or culture, or social life as organic unities superior to material or economic determinants and untroubled by inner fissures, dissent, or tension.

Orientalism A term given new meaning by Edward Said in his highly influential study of the same name (1979; first published 1978). Traditionally, an Orientalist was a scholar devoted to the study of 'the Orient' or the East. In this traditional usage the term had positive connotations – devotion to scholarship, a commitment to unearthing the secrets of another culture, and so on.

Said's development of the term draws out its concealed political allegiances – the fact that Orientalists were complicit with imperialism and that they effectively provided Europe with 'one of its deepest and most recurring images of the Other' (1979, 1; see the entry for OTHER). This statement suggests a certain Lacanian element in Said's development – or DECONSTRUCTION – of this term, but there are also Foucauldian elements too, in spite of Said's view of Foucault's own eurocentrism (see his essay on Foucault in Said 1984). The following comment certainly suggests a Foucauldian influence:

> . . . Orientalism can be discussed and analyzed as the corporate institution for dealing with the Orient – dealing with it by making statements about it, authorizing views of it, describing it, by teaching it, settling it, ruling over it: in short, Orientalism as a Western style for dominating, restructuring, and having authority over the Orient.
>
> . . .
>
> My contention is that without examining Orientalism as a discourse one cannot possibly understand the enormously systematic discipline by which European culture was able to manage – and even produce – the Orient politically, sociologically, militarily, ideologically, scientifically, and imaginatively during the post-Enlightenment period. (1979, 3)

Said is interested in Orientalism as a set of ideas that are regulated more to achieve internal coherence than to achieve 'any correspondence, or lack thereof, with a "real" Orient' (1979, 5). This is not, however, to say that Orientalism is merely a fantasy with no relationship to the real; Said insists that over many generations it involved 'considerable material investment', and that it serves as a sign of 'European-Atlantic power over the Orient' (1979, 6). Orientalism is, moreover, 'premissed upon exteriority' according to Said: that is to say, it always involves a non-oriental doing the studying; 'the Orientalist is outside the Orient, both as an existential and as a moral fact' (1979, 20, 21).

Other To characterize a person, group, or institution as 'other' is to place them outside the system of normality or CONVENTION to which one belongs oneself. Such processes of exclusion by categorization are thus central to certain IDEOLOGICAL mechanisms. If woman is other, then that which is particular to the experience of being a woman is irrelevant to 'how things are', to the defining conventions by which one lives. If members of a given racial group are collectively seen as other, then how they are treated is irrelevant to what humanity demands – because they are other and not human.

When granted a capital letter the term invokes Jacques Lacan's theory of the way in which the SUBJECT seeks confirmation of itself in the response of the Other. As he puts it one of his less opaque pronouncements in 'The Agency of the Letter in the Unconscious or Reason Since Freud',

> If I have said that the unconscious is the discourse of the Other (with a capital O), it is in order to indicate the beyond in which the recognition of desire is bound up with the desire for recognition.
>
> In other words this other is the Other that even my lie invokes as a guarantor of the truth in which it subsists. (1977, 172)

Also illuminating is Lacan's comment in 'On a Question Preliminary to any Possible Treatment of Psychosis', that the Other is, 'the locus from which the question of [the subject's] existence may be presented to him' (1977, 194).

According to Anthony Wilden,

> Lacan's Other represents the patrocentric ideology of our culture. The Other is only theoretically *ne-uter*, for it is not pure 'Otherness'. It is the principle of the locus of language and of the signifier, which for Lacan, is naturally the phallus . . . (1972, 261)

Without its capital letter, 'other' in Lacan's usage designates the other, imaginary self, which is first formed during the MIRROR STAGE when the infant confronts his or her own image.

See also *OBJET A / OBJET A*; STEREOTYPE.

Overcoding See CODE

Overdetermination From the work of Sigmund Freud: if a symbol is the result of several separate or related causes then it is described as overdetermined. Freud saw the dream symbol as overdetermined because its full explanation could not involve merely one source or MEANING but had to take account of several interrelated sources and meanings. Writing of one of his own dreams, Freud suggests that certain elements are to be seen as 'nodal points' 'upon which a great number of dream-thoughts converged, and [which] had several meanings in connection with the interpretation of the dream' (1976, 388). In the same passage, however, Freud goes on to suggest a slightly different meaning for the term: 'each of the elements of the dream's content turns out to have been "overdetermined" – to have been represented in the dream-thoughts many times over' (1976, 388–9).

The term achieved some fashionability in the 1960s following its use by the French Marxist philosopher Louis Althusser in his essay 'Contradiction and Overdetermination' (in Althusser 1969, 87–128). The use ushered in by Althusser was more historico-political: a range of different social forces could result in a single, overdetermined event such as a political revolution.

According to Althusser, he was 'not particularly taken' by the term, but used it 'in the absence of anything better, both as an *index* and as a *problem*, and also because it enables us to see clearly why we are dealing with something *quite different from the Hegelian contradiction*' (1969, 101). What Althusser seems most concerned with here is to use the concept as a means of arguing against a crude economic determinism; as he argues,

> [Overdetermination] is *universal*; the economic dialectic is never active *in the pure state*; in History, these instances, the superstructures etc. – are never seen to step respectfully aside when their work is done or, when the Time comes, as his pure phenomena, to scatter before His Majesty the Economy as he strides along the royal road of the Dialectic. From the first moment to the last, the lonely hour of the 'last instance' never comes. (1969, 113)

Thus overdetermination as concept served to warn against simplistic cause-and-effect views in a range of disciplines.

P

Paradigm shift. A term introduced by Thomas S. Kuhn in his *The Structure of Scientific Revolutions* (1970; first published 1962). Kuhn suggested that particular learned communities or specialities rested upon acceptance of 'a set of recurrent and quasi-standard illustrations of various theories in their conceptual, observational, and instrumental applications'. These, he proposed, are the community's *paradigms*, which can be found revealed in its 'textbooks, lectures, and laboratory exercises' (1970, 43). Kuhn (as his title suggests) was particularly interested in how changes take place in scientific thinking, and his concept of the paradigm plays a central rôle in his explanation. Kuhn's paradigms are not just the illustrations he mentions but also the assumptions which are to be found behind, and constituted by, these illustrations. In other words, a paradigm is constituted by a set of beliefs which both enables and constrains research: a framework or scaffold which can underpin or support further work but which of necessity also excludes a range of possibilities.

Readers of Kuhn's book from the Humanities were much struck by his accounts of cases of scientific evidence which was not recognized as such because it did not fit into known and accepted paradigms: for instance the scientist who actually isolated oxygen but was unable to recognize what he had done because it did not fit into the phlogiston theory within the confines of which he was working. This had more than a passing resemblance to then-influential theories of IDEOLOGY, and Kuhn's ideas attracted much attention outside the realms of the philosophy of science.

In particular, Kuhn's view of the necessity for a *paradigm shift* to enable major advances in scientific theory to take place seemed to have something (but only something) in common with what Louis Althusser was saying concerning the need for the theorist to move from ideology to science. The difference is that for Kuhn there is no promised land of science; paradigm succeeds paradigm like the succession of blinkered generational views with which Philip Larkin's poem 'High Windows' presents us, each seeming as if it represents an advance but each with its own inevitable limitations.

Kuhn's concept of the paradigm shift attracted criticism, however, because of its 'inwardness': the shift was engendered and triggered by the pressure of internal contradictions rather than by (as in traditional MARXIST views) the pressure of external forces which, in Darwinian mode, excluded theories which could not adapt to new external needs. The term has entered into literary-critical vocabulary in relation, very often, to arguments about the CANON (a WORK is only recognized as major when new literary paradigms allow it to be understood or appreciated), and in relation to arguments about interpretation (we interpret the evidence of a TEXT differently when our literary paradigms change, just as the scientist interprets the evidence provided by an experiment when his or her scientific paradigms change).

See also CONVENTIONALISM; COPERNICAN REVOLUTION; *ÉPISTÉMÈ*; EPISTEMOLOGICAL BREAK; PROBLEMATIC, for comparable concepts.

Paradigmatic See SYNTAGMATIC AND PARADIGMATIC

Paralanguage See PARATEXT

Paralepsis According to Gérard Genette, the presenting of more information in a narrative than is authorized by the attribution of focus (1980, 195).

Thus in Joseph Conrad's novel *Under Western Eyes* the NARRATOR (the 'Teacher of Languages') claims to base his account on the written report of the CHARACTER, Razumov. And yet after Razumov has finished this report and has wrapped it up, the Teacher of Languages continues to describe Razumov's actions and experiences.

Not to be confused with PARALIPSIS.

Paralipsis Gérard Genette suggests this term for an ELLIPSIS which is created not by missing out a temporal unit in the NARRATIVE succession, but by omitting a constituent element or elements in a period of STORY time that is covered in the narrative.

Not to be confused with PARALEPSIS.

Paraliterature See CANON

Paratext A coinage of Gérard Genette in his work *Seuils* (1987). The term is applied to those TEXTS which surround the text proper: prefaces, publishers' announcements, jacket descriptions, footnotes, and so on. The term is presumably applied on the model of *paralanguage*: paralinguistic features are those non-verbal accompaniments to a verbal utterance: grunts, voice pitch, facial expression, hand gestures and BODY movements, and so on.

See also PREPUBLICATION / POSTPUBLICATION READING.

Parergon See FRAME

Partial synonymy See SENSE AND REFERENCE

Patriarchy Technically, government by men – either within the family or in society at large – with authority descending through the father. In Lacanian theory, patriarchy is the internalization of the law of the father (see the entry for NAME-OF-THE-FATHER), which takes place when the individual enters human culture through the Oedipus Complex. In recent usage the term has been used to point both to the actual exercise of power and also to the IDEOLOGICAL system – the ideas and attitudes – used to bolster, justify, and protect this power. Patriarchy thus has political, economic, social and ideological dimensions.

Pause See DURATION

Performatives See SPEECH ACT THEORY

Perlocutionary act See SPEECH ACT THEORY

Personal See NARRATIVE SITUATION

Perspective and voice Gérard Genette has drawn attention to the importance of a long-neglected distinction between *who sees?* and *who speaks?*. In his terminology this distinction is expressed in terms of the opposition between mood and voice (1980, 30). According to Genette, the category of mood gathers together the problems of DISTANCE which American critics have traditionally discussed in terms of the opposition between *telling and showing*, and he suggests that these terms represent a resurgence of the Platonic terms DIEGESIS AND MIMESIS. *Voice*, in contrast, he reserves to describe the way the NARRATIVE SITUATION, along with the NARRATOR and his or her audience, is implicated in the narrative (1980, 29–31). To summarize: *mood* operates at the level of connections between STORY and narrative, while *voice* designates connections both between narrating and narrative, and narrating and story (1980, 32).

Thus, for example, although the narrative voice in Joseph Conrad's *Under Western Eyes* is that of the personified narrator, the English 'Teacher of Languages', what is observed (in both senses of the word) frequently goes beyond his actual consciousness, and the reader is told things which he could not possibly know if he were a living human being rather than a fictional character/narrator. (See the entry for PARALEPSIS.) Those following Genette have often preferred to replace *mood* by *perspective*, as the pair 'perspective and voice' matches up rather more neatly with 'who sees/who speaks'. In what follows I will adopt this terminology, and use 'perspective' to indicate 'who sees' and 'voice' to indicate 'who speaks'.

In Mieke Bal's terminology, perspective covers both the physical and psychological points of perception in a narrative, but it does not involve the actual *agent* performing the act of narration (1985, 101). Perspective can be further specified by means of another term used by both Genette and Bal: focalization. A narrative which has zero focalization is one in which it is impossible to fix the perspective in terms of which the narrated characters, events and situations are being observed and presented. Many works of fiction which were traditionally described as characterized by an *omniscient* point of view are described by modern narratologists as having a zero focalization. The term *focalized narrative*, in contrast, can also be rendered as *internal perspective*: the story is told from the perspective of a point (normally a consciousness) which is internal to the story, or intradiegetic. (Sometimes the term intrafictional is preferred.) Such focalized narratives can be *fixed*, as in Jean Rhys's *After Leaving Mr Mackenzie* where everything we learn comes from the heroine/narrator Julia Martin, *variable*, as in Emily Brontë's *Wuthering Heights* in which some events are presented from the perspective of Mr Lockwood, some from that of Mrs Dean, and some from that of Isabella (in her reported letter), or *multiple* as in Tobias Smollett's *Humphry Clinker* in which the epistolary technique allows the same events to be presented more than once from different and contrasting perspectives. The distinction between variable and multiple perspective is, clearly not always so very easy to establish.

External focalization denotes a focalization that is limited to what, were the story true, the observer could actually have observed 'from the outside'; in other words, it involves no accounts of characters' thoughts, feelings, and emotions unless these are revealed in external behaviour or admitted to by the characters. According to Genette, external focalization involves a perspective which is intradiegetic but outside that of the characters. In his Introduction to Genette's *Narrative Discourse*, Jonathan Culler has referred to the criticism made by Mieke Bal of Genette's distinction between *internal focalization* and *external focalization*. Culler notes that Bal claims that the two terms refer to rather different cases, for whereas

> In what Genette calls *internal focalization* the narrative is focused *through* the consciousness of a character, . . . *external focalization* is something altogether different: the narrative is focused *on* a character, not through him.
>
> (Genette 1980, 10)

An alternative way of referring to types of focalization is by means of reference to varieties of narrative *vision*.

Seymour Chatman has distanced himself from Genette's and Bal's use of the term 'focalization' and has suggested as alternatives 'slant' (to delimit 'mental activity on *this* side of the discourse-story barrier') and 'filter' (for 'capturing something of the mediating function of a character's consciousness', 'as events are experienced from a space within the story world' [1990, 144]).

Because of these variations in usage care needs to be exercised in the use of these terms.

See also ENUNCIATION; FRAME; NARRATIVE SITUATION; REFLECTOR (CHARACTER).

Petit récit See GRAND NARRATIVE

Phallic criticism See PHALLOCENTRISM

Phallocentrism A term dating from recent FEMINIST theory and used to refer to interlocking social and IDEOLOGICAL systems which accept and advance a PATRIARCHAL power symbolically represented by the phallus. 'Phallus' has to be understood as a *cultural* construction attributing symbolic power to the *biological* penis. The concept clearly builds upon artistic and CULTURAL representations of the penis which, in both ancient and modern societies, have been incorporated into ideological justifications of male power. (Jacques Derrida, incidentally, sees such a view of the phallus as an example of a transcendental signifier: see the entry for TRANSCENDENTAL SIGNIFIED.)

In contemporary feminist usage, phallocentric patterns of thought consciously or unconsciously assume and advance a view of the masculine as natural source of power and authority, and of the feminine as naturally subject to this. Thus man is PRESENCE, woman ABSENCE, and this absence is symbolically underwritten by overt or concealed reference to the absent phallus. For some feminists, the phallus has an even more all-embracing set of symbolic meanings; Madeleine Gagnon, for example, writes that

> The phallus, for me at this time, represents repressive capitalist ownership, the exploiting bourgeoisie, the higher knowledge that must be gotten over; it represents an erected France that watches, analyzes, sanctions. The phallus means everything that sets itself up as a mirror. Everything that erects itself as perfection. Everything that wants regimentation and representation. That which does not erase/efface but covets. That which lines things up in history museums. That which constantly pits itself against the power of immortality. (1980, 180)

The term seems to have entered contemporary feminism through the Lacanian critique of Freud. Lacan's essay 'The Signification of the Phallus', based upon a lecture delivered in 1958 and first published in French in 1966, is by no means feminist in orientation. But in exploring the phallus as signifier in this and other writings, Lacan opened the way to a feminist appropriation of the term which used, initially, a STRUCTURALIST conceptual framework.

Phallocratic denotes, literally, rule by the phallus – on the model of democratic, autocratic, and so on, and is used in feminist DISCOURSE to indicate a system of power and authority which is based upon phallocentric values.

Phallic criticism is criticism which abides by and furthers phallocentric values, which uses methods associated with the exercise of phallocentric power, and which is backed by the institutional resources of phallocentrism.

Phallocratic See PHALLOCENTRISM

Phallogocentrism A portmanteau term combining PHALLOCENTRISM and LOGOCENTRISM, coined by Jacques Derrida in his critique of Jacques Lacan in 'The Purveyor of Truth' (1975). For Derrida, Lacan's reading of Edgar Allan Poe's *The Purloined Letter* is guilty of phallogocentrism because Lacan sees the letter unproblematically as phallus, rather than recognizing that meaning cannot exist in such unproblematic one-to-one relationships.

Although the term was intended to imply criticism it has appealed to FEMINISTS and others eager to imply a connection between male, PATRIARCHAL authority, and systems of thought which LEGITIMIZE themselves by reference to some PRESENCE or point of authority prior to and outside of themselves. The hidden, legitimizing presence, in other words, is always, at root, that of the Father, whose authority is a starting point or unconsidered assumption rather than something that can be justified or admitted. The term implies that both phallocentrism and logocentrism have in common that they are both monolithic systems built round a single, ultimate determining CENTRE (the phallus, the word), a centre which ends indeterminacy and play and imposes meaning by the imposition of its unchallengeable authority.

David Lehman's explanatory note to a short poem by himself, published in the *Times Literary Supplement* for May 18–24, 1990, attributes to Benjamin Krull the following definition of phallogocentrism: 'what happens when you eliminate the space between the second and third words of the sentence *the pen is mightier than the sword*' (1990, 524). This suggests that the term's meaning has undergone a shift from that which Derrida originally intended (which would, presumably, neither surprise nor dismay Derrida).

See the useful extended discussion by Paul Julian Smith (1992).

Phatic See FUNCTIONS OF LANGUAGE

Phenomenology. Phenomenology originates in the writing of the German Edmund Husserl, whose philosophy takes as its starting point the world as experienced in our consciousness. It thus rejects the possibility of considering the world independently of human consciousness, but seeks rather to get back to concrete reality through our experience of it. For Husserl, consciousness is always consciousness of something: it is directed outwards rather than inwards – even if it is directed on to something imagined. It is thus too simple to describe phenomenology as idealist, for although it posits the impossibility of our gaining a knowledge of the world which is untouched by our perception of that world, it does suggest that through an eidetic method we can build up a successively more and more accurate understanding of the objects of our consciousness by filtering off accidental and personal elements in our perception of them. (Husserl takes the term *eidetic* from the Greek *eidos* meaning form, or type, and uses it to describe his method of abstracting universals from the flux of images given to us in consciousness. The method is supposed to be able to isolate what is constant and invariable in the objects of our consciousness, and can thus perhaps be used to remove the local and adventitious in our experience of a literary work.) In order so to analyse our consciousness we must suspend all preconceptions about the objects with which it is concerned (see the entry for *EPOCHÉ*). As Terry Eagleton comments,

> But if Husserl rejected empiricism, psychologism and the positivism of the natural sciences, he also considered himself to be breaking with the classical idealism of a thinker like Kant. Kant had been unable to solve the problem of how the mind can really know objects outside it at all; phenomenology, in claiming that what is given in pure perception is the very essence of things, hoped to surmount this scepticism. (1983, 56–7)

It is not hard to understand the interest that such ideas aroused amongst students of art and literature, for whom the pseudo-objectivity of positivism seemed no more applicable to the study and appreciation of art-works than did variants of Kantianism which too soon could end up in solipsism and an abandonment to unfettered personal 'taste'. A range of literary theorists and aestheticians found something in Husserl's ideas upon which they felt they could build. One of the earliest was the Polish aesthetician Roman Ingarden, who argued that a READING of a WORK of literature CONCRETIZES it (much as the PERFORMANCE of a play concretizes the written TEXT).

The Geneva School of criticism owes its largest debt to Husserl and phenomenological ideas. Its members were mostly associated with the University of Geneva, and have on occasions been referred to collectively as the 'Critics of Consciousness'. J. Hillis Miller, who was for a time closely identified with the school, has written a useful survey-account of the school which is reprinted in his book *Theory Then and Now* (1991). According to Hillis Miller, for the Geneva critics literature is a form of consciousness (1991, 14), while criticism is

> fundamentally the expression of a 'reciprocal transparency' of two minds, that of the critic and that of the author, but they differ in their conceptions of the nature of consciousness. From the religious idea of human existence in [Marcel] Raymond and [Albert] Béguin, to the notion in [Georges] Poulet's criticism that what counts most is 'the proof, the living proof, of the experience of inner spirituality as a positive reality,' [Jean] Rousset's belief that the artist's self-consciousness only comes into existence in the intimate structure of his work, the unquestioning acceptance of an overlapping of consciousness and the physical

> world in [Jean-Pierre] Richards's criticism, and the fluctuation between incarnation and detachment in the work of [Jean] Starobinski, these six critics base their interpretations of literature on a whole spectrum of dissimilar convictions about the human mind. (1991, 29)

This is criticism that is a long way away from the death of the author; the idea of writing as self-expression or self-projection is central to the work of Jean Rousset, and according to Hillis Miller, Georges Poulet believes that

> Criticism must therefore begin in an act of renunciation in which the critic empties his mind of its personal qualities so that it may coincide completely with the consciousness expressed in the words of the author. His essay will be the record of this coincidence. The 'intimacy' necessary for criticism, says Georges Poulet, 'is not possible unless the thought of the critic *becomes* the thought of the author criticized, unless it succeeds in re-feeling, in re-thinking, in re-imagining the author's thought from the inside . . . (1991, 15)

READER-RESPONSE CRITICISM owes an important debt to phenomenology and the Geneva School, particularly in the person of the German critic Wolfgang Iser. Iser's essay 'The Reading Process: a Phenomenological Approach' (reprinted in Iser 1974) is a good example of the creative development of a number of aspects of phenomenology – and, too, of the way in which phenomenology leads naturally to some of the preoccupations of reader-response critics. Take, for example, the first sentence of the essay:

> The phenomenological theory of art lays full stress on the idea that, in considering a literary work, one must take into account not only the actual text but also, and in equal measure, the actions involved in responding to that text. (1974, 274)

Iser then moves on to discuss Ingarden's theory of artistic concretization, and concludes that the literary work has two poles: the artistic pole (the text created by the AUTHOR), and the aesthetic pole (the realization accomplished by the READER) (1974, 274).

Iser stresses the work's *virtuality*: like the text of a play which can be produced in innumerable ways, a literary work can lead to innumerable reading experiences. (Compare Derrida's argument that 'reading is transformational' [1981b, 63].) He also makes use of Husserl's argument that consciousness is *intentional*, that it is directed and goal-seeking rather than random and all-absorbing. So far as the reading of literature is concerned, this allows Iser to place a high premium not just upon the reader's 'pre-intentions' – what he or she goes to the text with – but also upon the intentions awakened by the reading process itself (and, indirectly, by the text). One of the best-known of Iser's arguments involves the literary work's 'gaps'. According to him, no literary work is, as it were, complete: all have gaps which have to be filled in by the reader, and all readers and readings will fill these in differently (1974, 280; Iser makes similar points in other essays). Iser's gaps bear a close relationship to Roman Ingarden's 'spots of indeterminacy': see the entry for CONCRETIZATION.

See also SUB-TEXT; ABSENCE.

Phenotext See GENOTEXT AND PHENOTEXT

Phonocentrism See LOGOCENTRISM

Play See LUDISM

Pleasure Freud's *pleasure principle* dominated the new-born infant, but in the maturing or mature individual was placed under the sway of the *reality principle*, regaining its sway only in fantasy or day-dreaming (see the discussion in Mitchell 1974, 13). Pleasure was thus infantile, temporary, not to be confused with the adult experiences offered by literary WORKS – although to the extent that the READING of literature was seen to be regressive and comparable to dream-experiences then pleasure might enter into it.

However much common readers may have gone on using terms such as *pleasure* and *enjoyment* with reference to their reading of literary works, within modern movements in literary criticism during the present century these have generally had something of a thin time. In 1973, however, Roland Barthes's *Le Plaisir du Texte* was published (translated as *The Pleasure of the Text* in 1976), and pleasure suddenly became much more respectable amongst literary critics and theorists. Barthes's *plaisir* is difficult to translate into English, and has a generally more indulgent and more specifically sexual/sensual force than its more staid English counterpart. Barthes distinguishes between *pleasure of the text* and *texts of pleasure*: the former can be spoken, whereas the latter are perverse and are outside any imaginable finality (1976, 51–2). *The Pleasure of the Text* is, however, not an easy work to summarize or describe, and perhaps its most important effect was to bring the issue of reader-gratification through the reading experience back into the forefront of critical attention.

The issue of pleasure has also been taken up by recent FEMINIST theorists, some of whom have argued convincingly that pleasure has been NATURALIZED in particular ways within PATRIARCHAL societies, ways which devalue, MARGINALIZE or ignore specifically female forms of pleasure.

Poetic See FUNCTIONS OF LANGUAGE

Point of view See PERSPECTIVE AND VOICE

Politeness In a useful recent article entitled 'The Politeness of Literary Texts' Roger Sell traces the application of this term from 'the zenith of its lofty meaning' in the eighteenth century when it 'embraced intellectual enlightenment and civilization as prized by the Augustans', through to the way in which the term has been redefined by anthropological linguists (1991, 208). It is this most recent usage that has established the term's current process of redefinition: 'the strategies, linguistically realized, by which human beings hold their own or grasp for more' (Sell 1991, 211, after Brown & Levinson 1978 and 1987).

Sell details a number of recent attempts on the part of scholars interested in a *rapprochement* of Linguistics and Literary Studies to connect the concept of politeness with literary TEXTS, particularly with regard to the way in which these – or their AUTHORS – handle face-threatening acts (FTAs) in the language of authorial personae and CHARACTERS. (A face-threatening act is an act which challenges or endangers an individual's 'face', or public esteem. Thus to be bawled out by one's boss is one thing, but to have it done in front of one's colleagues threatens not just one's relation to him or her, but one's whole reputation and esteem. It is possible to criticize a person without threatening their face – for example, by pretending that an action was a mistake rather than done deliberately. Pragmatic skill normally involves avoiding either delivering or receiving FTAs whenever possible.)

Sell's own approach, however, is wider than that mentioned above: he sees all interaction, and all language, as operating within politeness parameters (1991, 215), a position that he associates with Mikhail Bakhtin's view that all language use is DIALOGIC, and thus to some extent engages with those others who hear or read it. For Sell politeness is 'a matter of choice and co-operativeness in interpersonal relations' (1991, 221), more a form of helpfulness than the purely selfish manoeuvring for personal advantage some other theorists view it as. He distinguishes between *selectional* and *presentational* politeness, the former involving the avoidance of anything 'threatening the readers' positive or negative face' and the latter observing 'the co-operative principle at all costs, so that his readers would never be in the slightest doubt as to what was happening, what he meant, or why he was saying what he was saying' (1991, 221, 222).

As this makes clear, the concept owes a lot to, or has much in common with, aspects of both SPEECH ACT THEORY and PRAGMATICS. It can be added that once one starts to talk about the politeness of literary texts the way is open to talk about their impoliteness. Just as impoliteness can be a technique for getting one's own way in interpersonal relations, so too an author may challenge and manipulate a reader's expectations and responses through, for example, making it very difficult for him or her to understand what is happening, what is meant, or why what is being said is being said. Much early MODERNIST literature can be said to have recourse to impoliteness of this sort.

Polyglossia See HETEROGLOSSIA

Polyphonic Literally, many-voiced. The term is associated with the theories of Mikhail Bakhtin, especially with his idea of the *polyphonic novel*. In his study of Dostoevsky Bakhtin argues that Dostoevsky was the creator of the polyphonic novel:

> *A plurality of independent and unmerged voices and consciousnesses, a genuine polyphony of fully valid voices is in fact the chief characteristic of Dostoevsky's novels.* What unfolds in his works is not a multitude of characters and fates in a single objective world, illuminated by a single authorial consciousness; rather a *plurality of consciousnesses, with equal rights and each with its own world,* combine but are not merged in the unity of the event. (1984, 6)

Bakhtin uses the word *voice* in a rather special way, to include not just matters linguistic but also matters relating to IDEOLOGY and power in society. A voice for Dostoevsky refers the listener not just to an originating person, but to a network of beliefs and power relationships which attempt to place and situate the listener in certain ways. As he puts it,

> Language is not a neutral medium that passes freely and easily into the private property of the speaker's intentions; it is populated – overpopulated – with the intentions of others. Expropriating it, forcing it to submit to one's own intentions and accents, is a difficult and complicated process. (1981, 294)

That process is the means whereby *language* is transformed into a *voice*.

Bakhtin's view that the polyphonic novel was born with Dostoevsky has been challenged, and he himself seems sometimes to have accepted that polyphonic elements can be found in many novels in many CULTURES and times. But his work on Dostoevsky has drawn attention to a representative lack of a single CENTRE of authority in the modern (or MODERNIST) literary WORK. And this can be said to relate to arguments about the death of the AUTHOR.

Brian McHale points out that a rather different use of the concept can be found in the work of the Polish PHENOMENOLOGIST Roman Ingarden. According to Ingarden,

> the literary artwork is not ontologically uniform or monolithic, but *polyphonic*, stratified. Each of its layers has a somewhat different ontological status, and functions somewhat differently in the ontological make-up of the whole.
> (McHale 1987, 30)

For Ingarden there are four such strata: that of word-sounds, that of meaning-units, that of 'presented objects' (which are distinguished from 'real-world objects'), and that of 'schematized aspects' ('presented objects' do not have the same determinacy as 'real-world objects', and they are inevitably schematic – which is why it bespeaks a failure to approach the artwork as artwork if one asks questions such as 'How many children had Lady Macbeth?') (based on McHale 1987, 33).

See also DIALOGIC; HETEROGLOSSIA.

Polyvalent discourse See REGISTER

Polyvocality See POLYGLOSSIA

Popular Dissatisfactions with the literary CANON, along with the impact of interdisciplinary ventures involving not just literature but also other art-forms and the products of the mass media, have led to a renewed interest in the issues raised by the word *popular*. Raymond Williams points out that the word stems from a legal and political term meaning 'belonging to the people', but that in the historical shifts towards the widespread modern usage of 'well-liked' there has been a strong pejorative streak relating to the implication the word carries of 'setting out to gain favour' (1976, 198).

In current critical usage terms such as *popular culture*, *popular literature* and *popular art* typically cover the senses both of '*for* the people' and '*of* the people'. Williams suggests that this is a relatively recent development, as previously a term such as 'folk culture' was reserved for usages involving the culture produced by people for themselves. We can suggest, however, that the etymological relation between *popular* and *people* inevitably leads the former word to respond to shifts of attitudes and meaning attached to the latter one – especially of a socio-political nature. Thus during 1938 Bertolt Brecht's polemics against the anti-MODERNIST positions of Georg Lukács included interesting proposed definitions (working definitions in the sense that they were intended to be put to work in specific cultural practices) of both *popular* and CULTURE. The link between views of 'the people' (both the term and the reality to which it refers) and views of the *popular* are clear in Brecht's comments:

> Our concept of what is popular refers to a people who not only play a full part in historical development but actively usurp it, force its pace, determine its direction. We have a people in mind who make history, change the world and themselves. We have in mind a fighting people and therefore an aggressive concept of what is *popular*.
>
> Popular means: intelligible to the broad masses, adopting and enriching their forms of expression / assuming their standpoint, confirming and correcting it / representing the most progressive section of the people so that it can assume leadership, and therefore intelligible to other sections of the people as well / relating to traditions and developing them / communicating to that portion of the people which strives for leadership the achievements of the section that at present rules the nation. (Bloch *et al.* 1977, 81)

The very considerable influence of the work of Mikhail Bakhtin during the 1980s also involved a renewal of interest in the concept of the popular, directing attention towards traditions of expression and resistance both within and outside canonical literature and art.

See also CULTURE; DIALOGUE; IDEOLOGY.

Popular culture See CULTURE; POPULAR

Pornoglossia According to Deborah Cameron, 'since the word *pornography* means "pictures of prostitutes", perhaps *pornoglossia* would be a good name for the language which reduces all women to men's sexual servants' (1985, 77). Pornoglossia, then, is a use of language to describe women purely in terms of their sexual usefulness, availability, or attractiveness to men.

Position / positionality See READING POSITION

Postmodernism See MODERNISM AND POSTMODERNISM

Post-structuralism A term that is sometimes used almost interchangeably with DECONSTRUCTION, while at other times being seen as a more general, umbrella term which describes a movement of which one important element is deconstruction. Thus Richard Harland, for example, suggests that the post-structuralists fall into three main groups: the *Tel Quel* (a French journal) group of Jacques Derrida, Julia Kristeva and the later Roland Barthes; Gilles Deleuze and Félix Guattari (authors of the influential *Anti-Oedipus: Capitalism and Schizophrenia* [published in French in 1972]) and the later Michel Foucault; and (on his own) Jean Baudrillard (Harland 1987, 2). Whether Jacques Lacan is a structuralist and/or a post-structuralist is a matter for continuing debate.

The degree of uncertainty that surrounds the use of this term can, however, be suggested by noting that Alex Callinicos proposes a rather different division of post-structuralism into two main strands of thought. The first of these is what Richard Rorty has dubbed TEXTUALISM, while the second is one in which the master category is Michel Foucault's 'power-knowledge'. This 'worldly post-structuralism', as Callinicos calls it using a term of Edward Said's, involves an 'articulation of "the said and the unsaid", of the discursive and the non-discursive' (Callinicos 1989, 68). Callinicos argues that whereas the textualists see us as imprisoned in TEXTS, unable to escape the discursive (or unable to see any reality unmediated by DISCOURSES), 'worldly post-structuralism' leaves open the possibility of contact with a reality unmediated by or through discourses.

If one accepts this division it has to be said that post-structuralism in its textualist version has had a far more significant impact upon Literary Studies than has the Foucauldian variant, although many of Foucault's ideas have been taken up for criticism and development by FEMINIST critics.

'Textualist' post-structuralism represents at the same time both a development and a deconstruction of STRUCTURALISM – a demonstration of its argued inner contradictions. A classic example of this is to be found in an early (1968) interview of Derrida by Julia Kristeva, published in *Positions*. Derrida here takes issue with what he claims is Saussure's maintenance of a rigorous distinction between 'the *signans* and the *signatum*, [and] the equation of the *signatum* and the concept', which, he argues,

> inherently leaves open the possibility of thinking a *concept signified in and of itself*, a concept simply present for thought, independent of a relationship to language, that is of a relationship to a system of signifiers . . . (1981b, 19)

Within Saussure's revolutionary view of language as a system of DIFFERENCES with no positive forms, that is, Derrida argues that one can discover (by deconstructing Saussure's argument) a relic of the old ideas, an extra-systemic entity, a TRANSCENDENTAL SIGNIFIED. By pursuing the implications of Saussure's arguments as far as possible one is able to go beyond them: the rigorous structuralist thus stands like Keats's Cortez, staring at the Pacific of post-structuralism (although his wild surmise is hardly silent).

Derrida's own position is best followed through certain key terms such as deconstruction, LOGOCENTRISM, DIFFÉRANCE, transcendental signified, METAPHYSICS OF PRESENCE, and so on – which to avoid repetition I will not reconsider here. Suffice to say that central to his endeavour (as he himself admits) has been a commitment to the rooting out of a belief in absolute and extra-systemic determinants of meaning.

Power In NARRATIVE theory, that which either allows or prevents a SUBJECT from reaching its object. A power can be an individual CHARACTER, or it can be an abstraction such as fate, age, nature, and so on (see Bal 1985, 28).

The issue of power has entered into recent discussion of literature mainly in connection with issues loosely related to IDEOLOGY. Literature is often subject to control (through censorship, restrictions on literacy, use of powers associated with the ownership of libraries, publishing houses, reviewing media) precisely because it can challenge existing authorities – or because these at least believe that it can.

Pragmatics From two important early theorists of SEMIOTICS, Charles Morris and Charles Peirce, has come a useful tripartite distinction between *syntactics* (SIGNS and their relations to other signs), *semantics* (signs and their relations to the 'outside world'), and *pragmatics* (signs and their relations to users). The distinction has been most fruitful within the realm of Linguistics, where it has enabled theorists such as Ferdinand de Saussure and those who followed his example to isolate different systems of formal rules (syntactic and semantic) from language in its actual, day-to-day use (its pragmatic existence). Saussure and his disciples argued that language at the pragmatic level was subject to too many random and unquantifiable pressures to constitute a proper object of study. To be studied it had to be reduced to an ideal state untouched by the accidents and casual pressures of everyday usage.

Ignoring the pragmatic aspect of language use in this way was very useful to linguisticians, and during the time of Saussure's greatest influence was seen by some to be a precondition for the formal study of language. At the same time, during the 1960s and 1970s, the theories of the American linguist Noam Chomsky also argued for a linguistic object of study which could be isolated apart from the pragmatic use of language.

But in recent years more and more disquiet has been expressed at the banishing of the realm of the pragmatic from the concern of the theorist of language, and pragmatics, which has its primary origin within the discipline of

Linguistics, involves a certain reaction against the ideas and practices of Saussure and Chomsky. Thus Stephen Levinson, in his *Pragmatics*, suggests that the recent growth of interest in pragmatics owes much to a reaction against Chomsky's treatment of language as an abstract device or mental ability which is dissociable from the uses, users, and functions of language (1983, 35). He adds, however, that this is not the whole story:

> Another powerful and general motivation for the interest in pragmatics is the growing realization that there is a very substantial gap between current linguistic theories of language and accounts of linguistic communication. . . . For it is becoming increasingly clear that a semantic theory alone can give us only a proportion, and perhaps only a small if essential proportion, of a general account of language understanding. (1983, 38)

This development in Linguistics has its parallels in other disciplines, including Literary Criticism. A recent collection of essays edited by Roger Sell and entitled *Literary Pragmatics* contains a number of attempts to transpose some of the more general principles of pragmatics to a literary context. Central to such a project is a commitment to moving away from the study of literary WORKS as closed or purely formal structures of TEXT to a recognition of them as MEDIATING elements in chains of communication.

> Literary pragmatics takes for granted that no account of communication in general will be complete without an account of literature and its contextualization, and that no account of literature will be complete without an account of its use of the communicative resources generally available. In effect, it reinstates the ancient linkage between rhetoric and poetics . . . (1991, xiv)

Literary pragmatics typically tries to bring together CENTRIFUGAL AND CENTRIPETAL movements: moving into the text and isolating pragmatic techniques (implicature, presupposition, persuasion), and relating these to forces outside the text in the worlds of writer and reader such as power relations, CULTURAL traditions, systems of publishing and distribution, censorship, and so on, with a stress throughout on *particular* pragmatic conjunctions and interactions.

Not to be confused with Pragmatism, a philosophical doctrine which recommends a concentration upon that which is of practical concern for human beings and which avoids the consideration of more absolute or abstract issues.

See also DISCOURSE.

Prague School (Alternatively *Prague [Linguistic] Circle*.) A group of theorists writing in Prague from the late 1920s through to the German invasion of Czechoslovakia and a little beyond, some of whom were emigrés from the Soviet Union (the best-known amongst whom was Roman Jakobson, who moved to Prague in 1920 and was a founder member of the group at its inception in 1926). The best-known members of the group, apart from Jakobson, were René Wellek, Felix Vodička and Jan Mukařovský.

According to many commentators the Prague School writers constitute a clear bridge between RUSSIAN FORMALISM and modern STRUCTURALISM – a claim which can be confusing to the uninitiated as English translations of Prague School writings often have their authors describing their own position as structuralist. The structuralism of the 1960s and later is, however, somewhat (but not completely) different from the structuralism of the Prague School. What the two have in common is a significant debt to the writings of Ferdinand de Saussure, and an attempt to extend the application of Saussure's theories beyond language. The sub-title of *A Prague School Reader*, Paul L. Garvin's useful collection of some of the more important non-linguistic Prague School texts, 'On Esthetics, Structure, and Style', is a useful indication of the extent to which their work moved beyond Linguistics as such (see Garvin 1964). The starting point may be Saussure and language, but the work of the Russian Formalists, upon which the Prague writers built, is at least as important a source as is the work of Saussure. Much of the work of the group had a specifically linguistic focus – working with phonetics, phonology and semantics and attempting to define such concepts as 'phoneme', 'distinctive/redundant feature', and so on, but from this base the members of the group moved into a number of cognate areas, including those of literature and aesthetics. But none of this work was untouched by formalism and formalist ideas: Mary Louise Pratt argues that although the School's members made a point of calling their linguistics 'functional', like Saussure they were concerned almost uniquely 'with the function of elements within the linguistic system rather than with the functions the language serves within the speech community' (1977, 7) – although she adds that in the pre-Second World War period the poeticians of the group *were* concerned with this latter issue – as can, incidentally, be seen from the essays in Garvin's *Reader*. This, she adds, meant that from the publication of the important Prague School Theses of 1929, an opposition between poetic and non-poetic language is built into the group's pronouncements on the social function of language. The position is clearly expressed in Jan Mukařovský's essay 'Standard Language and Poetic Language' (reprinted in Garvin 1964).

With the movement of René Wellek to the United States an important link – between the Prague School and Anglo-American New Criticism – can be traced. *Theory of Literature* by René Wellek and Austin Warren first appears in the United States in 1949, at the beginning of the period of the New Criticism's greatest influence (perhaps a more accurate word would be HEGEMONY) in that country. It is fascinating to read Felix Vodička's 1942 essay 'The History of the Echo of Literary Works' in conjunction with W. K. Wimsatt's and Monroe Beardsley's highly influential 'The Intentional Fallacy' and 'The Affective Fallacy'. Two brief quotations from Vodička's essay must suffice to make this point:

> The literary work, upon being published or spread, becomes the property of the public, which approaches it with the artistic feeling of the time.
>
> . . .
>
> Subjective elements of valuation, stemming from the momentary state of mind of the reader or his personal likes and dislikes, must in the historical criticism of sources be separated from the attitude of the times, because the object of our cognition are those features which have the character of historic generality.
>
> (1964, 71)

The development of the Russian Formalist concept of DEFAMILIARIZATION into the similar concept of FOREGROUNDING represents perhaps the most influential of Prague School contributions to literary-critical theory, especially within the fields of NARRATOLOGY and STYLISTICS.

Prefiguring See PROLEPSIS

Prepublication / postpublication reading Adopted by Jerome J. McGann from John Sutherland's account of,

> on the one hand, the 'English' model of postpublication reading via the traditional reviewing institutions; and, on the other, the 'American' model of prepublication reading via early promotional apparatuses, advance printing and purchase orders, and various peripheral schemes for disseminating (not necessarily the text but) the presence of the text. (McGann 1991, 125)

McGann's adoption of the terms is in line with his interest in the way in which extra-textual factors affect the READING and reception of a literary WORK.

See also PARATEXT.

Presence According to Jacques Derrida in the early and influential article 'Structure, Sign and Play in the Discourse of the Human Sciences', 'all the names related to fundamentals, to principles, or to the center have always designated an invariable presence – *eidos, archē, telos, energeia, ousia* (essence, existence, substance, subject) *alētheia*, transcendentality, consciousness, God, man, and so forth' (1978, 279–80). All of these represent extra-systemic entities, points of reference or CENTRES of authority which escape from that play of DIFFERENCE which, following Saussure, Derrida believes to be the sole source of MEANING. The *metaphysics of presence*, then, is a LOGOCENTRIC belief in and reliance upon some such extra-systemic point of authority.

See also CENTRE; COPERNICAN REVOLUTION.

Presentism On the model of racism, sexism etc.: the imposition of the standards, values and attitudes of the present on the past (and, on occasions, the future). Presentism universalizes the present and refuses to perceive other times in their particularity and their movement.

Primal scene See PRIMARY PROCESS

Primary process In Sigmund Freud's *The Interpretation of Dreams* Freud posits the existence of two psychical forces (or currents, or systems), one of which constructs the wish which is expressed by the dream whilst the other exercises a censorship upon the dream-wish and forcibly brings about a distortion of its expression (1976, 225). These two forces have come to be known as the primary and the secondary processes in discussion of Freud's theories.

Juliet Mitchell has glossed the primary process as 'the laws that govern the workings of the unconscious' (1974, 8) and the primary process is characterized by the free and unrestrained nature of its operation. Of interest to literary critics has been the suggestion that the primary process has a significant or dominating responsibility with regard to artistic and literary composition.

Not to be confused with the *primal scene*, which in Freudian terminology is the child's witnessing of its parents' copulation.

Privilege The verb *to privilege* and cognate noun and adjective forms are frequently used by recent theorists to indicate a hierarchical structuring of causal factors. To privilege gender in a consideration of class and gender in historical change, then, is to suggest that gender is more influential than class in historical causality.

When NARRATIVE theorists talk of privilege, however, they generally refer to the possession of information which is either exclusive to a particular narrator or to some form of understanding which is not shared by the CHARACTERS (or other characters) in the narrative.

Proairetic code See CODE

Problematic Problematic as noun came into the vocabulary of Anglo-Saxon theorists through Louis Althusser's *For Marx*, first published in English translation in 1969. According to Althusser he borrowed the concept (in French, of course) from Jacques Martin, 'to designate the particular unity of a theoretical formation and hence the location to be assigned to this specific difference' (1969, 32).

Although Althusser assigned the term to *theoretical* formations it is often used to designate IDEOLOGICAL ones as well; any complex of beliefs which (whatever their implicit or explicit contradictions) hang together in a self-supporting unity may be referred to as a problematic by recent theorists. It is probably too late to alter this drift of meaning, but Althusser's usage does have the virtue of fitting in with his concept of EPISTEMOLOGICAL BREAK, that is, of a shift from one problematic to another such as he claimed to discover in the writings of Karl Marx.

The term can be used in a way that makes it appear similar to Foucault's concept of the *ÉPISTÉMÈ*, in as much as it may be believed that a particular problematic represents what is 'thinkable' for those in its grip, and beyond which thought cannot venture.

See also PARADIGM SHIFT for another comparable concept.

Projection See PROJECTION CHARACTERS

Projection characters CHARACTERS into whom an author projects (often contradictory) aspects of him or herself. The term *projection* is taken from the work of Sigmund Freud, who used it to describe a process in which impulses or beliefs unacceptable to the ego are attributed to someone else.

The opposite of projection is *introjection*, in which an individual as it were internalizes things from outside him/herself and looks on them as his or her own. According to Nicola Diamond (1992, 177), the term was coined by Sandor Ferenczi in 1909 (see Ferenczi 1909), but was adopted and developed by Freud. The process described by the term has interesting points of contact with claims concerning the internalization of the VOICES of others made by members of the Bakhtin circle and by the Russian L. S. Vygotsky (1986). (See the entry for INTERIOR DIALOGUE.)

Prolepsis Also, following Prince (1988), *anticipation*, *flashforward* or *prospection*. Any narrating of a narrative EVENT before the time in the STORY at which it will take place has been reached in the NARRATIVE. Some theorists, including Genette, have argued that the term be extended to the *evocation* (as well as the narrating) of such 'future' events. If this is accepted, it will follow that many prolepses cannot be recognized as such on first reading of a narrative: the story of the 'gringos' who disappear looking for silver at the opening of Joseph Conrad's *Nostromo* is proleptic of the final fate of many other CHARACTERS in the novel, but the evocation of these fates by this early account can take place only in the READER's memory or upon a second reading of the novel. Prolepses would thus be either overt or implied. It is also possible to use terms such as *prefiguring* and *foreshadowing* for the evocation of future events, and to reserve prolepsis for more overt narrative references to such events.

It should be added that even overt prolepses may be difficult to recognize as such on first reading of a narrative; they may initially be read as examples of ACHRONICITY.

A *completing prolepsis* is one which is needed to achieve a full chronological coverage because of a later omission in the narrative; a *repeating prolepsis* or *advance notice* supplies information that will be provided afresh later on in the narrative.

Gérard Genette distinguishes between *internal prolepses*, which are sited within the time-span of the story, and *external prolepses*, which are sited outside of the story's temporal limits (1980, 68–71).

Prospection See PROLEPSIS

Psycho-narration See FREE INDIRECT DISCOURSE

Punctual anachrony See ANACHRONY

Punctuation Punctuation in its traditional sense has been of considerable concern to literary scholars and critics; the task of establishing a reliable TEXT has always involved determining its punctuation – often no easy job, as writing and publishing CONVENTIONS regarding punctuation are not fixed or stable.

More recent theorists in a range of different disciplines have given a metaphorical meaning to this term, extending its coverage to include more than just the insertion of stops or divisions in written (or spoken) language for the purpose of refining MEANING and limiting ambiguity. Social psychologists, for example, have drawn attention to the way in which conflicting views of the same process of interpersonal process can be explained by reference to the manner in which the participants involved punctuate the interaction differently. (For example: 'I drink because she nags me'; 'I nag him because he drinks'.)

Punctuation, in other words, involves the imposition of order and causality on a set of otherwise discrete facts by means of boundary-setting and grouping. It is not just the syntax of sentences that can be established by punctuation, but the syntax of relationships – and of literary WORKS. AUTHORS can impose a certain amount of punctuation on their works by means of chapter and other divisions: anyone who feels that this is not the case should read a work such as Joseph Conrad's *The Shadow-Line* both in its manuscript form (in which there are no divisions) and in its published version (in which the work is divided into numbered sections).

In an original essay on James Joyce's *A Portrait of the Artist as a Young Man*, Maud Ellmann has devoted considerable attention to the issue of punctuation in the work, showing how the growth and maturing of the hero, Stephen, involves experimentation with punctuation activities, and that these are to be seen both as attempts to master the world and as the unwitting imposition of intermittency upon himself (1981, 197–8).

Jacques Derrida's term *espacement*, normally translated as *spacing*, seems to mean something very similar to punctuation in its broader sense.

The term *segmentation* carries a comparable force within Linguistics, where it refers to the way in which language is 'chunked' or divided into segments. Segmentation is most frequently discussed at the level of intonation, and it is well-known that a change in intonation can alter the meaning of an utterance. (Some of the humour in Peter Sellers's parodic version of the Beatles' *A Hard Day's Night*, which imitates Sir Laurence Olivier's 'Richard the Third' delivery, depends upon re-segmentation of the well-known lyrics.) Segmentation can also, however, be applied to written language. Where written language is concerned it is clear that segmentation and punctuation (in the traditional sense) are not necessarily synonymous. Punctuation MARKS may be taken as the visual indicators of a sort of internal or ideal intonation, but they are not just that nor do they exert complete control over such sense-determining 'intonation'.

According to Geoffrey Leech and Michael Short, segmentation is, along with sequence and salience, one of the three main factors of textual organization (1981, 217). (Salience involves 'significant prominence' – for example, as caused by 'end focus', or the principle that we notice that which comes last in a sequence.) For sequence, see the entry for FUNCTION.

Q

Quality (maxim of) See SPEECH ACT THEORY

Quantity (maxim of) See SPEECH ACT THEORY

Quest narrative In Anne Cranny-Francis's definition, a linear NARRATIVE in which temporal sequence is taken to signify material causation. She sees this as the dominant structure in nineteenth- and twentieth-century Western writing, a structure which carries a particular IDEOLOGICAL and political charge, for it PRIVILEGES linear sequence rather than, say, seasonal cycles, and thus underwrites DOMINANT DISCOURSES in our CULTURE (1990, 10, 11).

There are significant differences in the way in which heroes and heroines experience the quest in traditional (non-FEMINIST) narrative. The hero is active in his quest, overcoming difficulties and aided by helpers, whereas the heroine is passive (is often the object of quest), has to endure suffering and humiliation, and rather than overcoming the world has to learn to adapt to its demands (as these are MEDIATED by men).

R

Radial reading See READERS AND READING

Radical alterity A term used to describe that belief in the essential self-ALIENATION of the SIGNIFIER from itself that can be found in the writing of Jacques Derrida. The following comment illustrates the concept.

> The *representamen* [which we can render as *signifier*] functions only by giving rise to an *interpretant* that itself becomes a sign and so on to infinity. The self-identity of the signified conceals itself unceasingly and is always on the move. The property of the *representamen* is to be itself and another, to be produced as a structure of reference, to be separated from itself. (1976, 49–50)

In other words, signifiers are never themselves alone. Derrida criticizes Saussure's Linguistics for concealing a self-identical signified within what, on the surface, appears to be a relational system based on DIFFÉRANCE. In fact, he claims, behind this overtly differential system lies the concealed PRESENCE of a signifier the identity of which is prior to différance. In opposition to this he argues for the recognition of the signifier's radical alterity – or essential 'difference-from-itself'.

Derrida's use of the term *exteriority* invokes many of the same issues as does the term radical alterity. According to Derrida,

> The exteriority of the signifier is the exteriority of writing in general, and . . . there is no linguistic sign before writing. Without that exteriority, the very idea of the sign falls into decay.
>
> . . .
>
> Thus, within this epoch, reading and writing, the production or interpretation of signs, the text in general as fabric of signs, allow themselves to be confined within secondariness. (1976, 14)

See also EXTERIORITY for Michel Foucault's use of this term.

See also SIGN.

Radical feminism See FEMINISM

Radicle See ROOT / RADICLE / RHIZOME

Reach See ANALEPSIS

Readerly and writerly texts Translated from Roland Barthes's coinages *lisible* and *scriptible* in his *S/Z* (1990, 4–5). Barthes uses the terms to distinguish between traditional literary WORKS such as the classical novel, with their reliance upon CONVENTIONS shared by writer and READER and their resultant (partial) fixity or CLOSURE of meaning, and those works produced especially in the present century which violate such conventions and thus force the reader to work to produce a meaning or meanings which are inevitably other than final or 'correct'. Thus

> The writerly text is a perpetual present, upon which no *consequent* language (which would inevitably make it past) can be superimposed; the writerly text is *ourselves writing*, before the infinite play of the world (the world as function) is traversed, intersected, stopped, plasticized by some singular system (Ideology, Genus, Criticism) which reduces the plurality of entrances, the opening of networks, the infinity of languages. (1990, 5)

Readerly texts, in contrast, are products rather than productions, and they make up the enormous mass of our literature (1990, 5).

Behind these comments lies a polemical element: Barthes, as he himself admits, wishes to challenge the conventional division of labour between producers and consumers, between writers and readers. The writerly text involves such a challenge, for it forces the reader to engage in the process of writing – that is, it forces him or her to be creative in the manner traditionally limited to the writer-function. '[T]he goal of literary work (of literature as work) is to make the reader no longer a consumer, but a producer of the text' (1990, 4). This of course ties in with Barthes's comments on the death of the author – see the entry for AUTHOR.

A comparable distinction is that provided by Umberto Eco in his *The Role of the Reader* (1981) between OPEN AND CLOSED TEXTS.

Reader-response criticism See READERS AND READING

Readers and reading According to the American critic Stanley Fish, in a book published in 1980:

> Twenty years ago one of the things that literary critics didn't do was talk about the reader, at least in a way that made his experience the focus of the critical act. (1980, 344)

Since the time about which Fish is writing, however, more and more attention has been devoted to the identity, rôle and function of readers of literature, and this has led to the coining of a range of new terms. Some of this attention has stemmed from a number of different critical theories and approaches which are often collectively described as *reader-response criticism*, but some of it can best be seen in relation to the loosening of New Critical dogmas. Once appeal to the intentional and affective fallacies lost its force, curiosity about the reader faced no interdiction. A sure indication of this sense of new-found freedom is the gradual movement away from talk of 'the reader' towards reference to 'readers' – a movement accompanied by the willingness of critics to say 'I' rather than 'we', which in its turn has to be related to a growing awareness of the dangers of ethnocentricity and parallel GENDER and class biases. For this reason it may be a little misleading to think of reader-response criticism as a school: the term gathers together a range of attempts to theorize about readers and to study them and the reading process, and Terry Eagleton's joke description 'The Reader's Liberation Front' catches some of the diversity in the reality described. Indeed, not all criticism categorized as reader-response criticism is actually concerned with readers' response(s); much of it is concerned with other issues: readers' COMPETENCE, the reading process *in toto*, the TEXT's formation of the reader, and so on.

Perhaps not surprisingly, much of the criticism described as reader-response criticism is concerned with the novel. We are much more conscious of reading as a process when reading a novel than when reading a poem – especially a lyric poem which can be held as a finished unit more easily in the memory.

Wayne Booth's 1961 book *Rhetoric of Fiction* popularized the notion of the IMPLIED AUTHOR, and by extension the term implied reader was coined to describe the reader which the text (or the author through the text) suggests that it expects (Booth himself talks of the *postulated* or *mock* reader). It should be noted that this term, like all other terms with the form 'the X reader', although singular, actually has the purpose of describing a group or category of readers. Jacques Derrida has argued that all reading is transformational, and in the same comment warned against reading (in this case the classic texts of MARXISM) 'according to a hermeneutical or exegetical method which would seek out a finished signified beneath a textual surface' (1981b, 63). Although Derrida is not normally thought of as a reader-response critic, the same insistence upon those transformational and creative aspects of reading which we associate with this group of critics can be found in his work.

Closely related to the implied reader is the *inscribed reader*, that is, the reader whose characteristics are actually there to be discovered in the text itself, waiting for the actual reader to slip on like a suit of clothes. Umberto Eco has introduced the similar concept of the *model reader*; he argues that

> To make his text communicative, the author has to assume that the ensemble of codes he relies upon is the same as that shared by his possible reader (hereafter Model Reader) supposedly able to deal interpretatively with the expressions in the same way as the author deals generatively with them. (1981, 7)

This may be taken to imply that the model reader is external to the text, but later in the same chapter of his book Eco makes it clear that for him the concept is more intra-textual in nature and that the model reader is thus also to be understood to be inscribed in the text:

> In other words, the Model Reader is a textually established set of felicity conditions . . . to be met in order to have a macro-speech act (such as a text is) fully actualized. (1981, 11; for *felicity conditions* see SPEECH ACT THEORY)

Rather different is the *intended reader*, because here the evidence may be either intra- or extra-textual: an author's comment in a letter that a WORK of literature was written to be read by a particular person or group of people can be used as evidence substantiating a case for a particular intended reader, but clearly not for a particular inscribed reader. Again related but slightly different are the *average* and the *optimal* or *ideal reader* (sometimes translated as *super-reader*). Super-reader comes from Michael Riffaterre (he later replaces it with *archi-lecteur* or composite reader), and describes (paradoxically) as much readings as readers, in other words, the responses engendered in different readers by particular textual elements. Riffaterre has also coined the term *retroactive reading* to describe a second-stage, HERMENEUTIC reading which comes subsequent to an initial, *heuristic reading* or reading-for-the-meaning process (1978, 5; see the entry for MEANING AND SIGNIFICANCE).

Different again is *symptomatic reading*: if we define a symptom as a non-intentional SIGN then it follows that a symptomatic reading treats a work much as a doctor examines a patient for symptoms. The doctor (in the exemplary situation) does not ask the patient what is wrong with him or her, but looks for clues the meaning of which the patient is ignorant. A symptomatic reading, then, seeks to use such clues in a work as a way into the secrets of the author, or of his or her society or CULTURE, or whatever.

Jerome J. McGann has coined the term *radial reading* for the sort of reading that 'puts one in a position to respond actively to the text's own (often secret) discursive acts'. As an example he refers to the two volumes which comprise the first edition of Ezra Pound's first twenty-seven cantos, each of which 'makes a clear historical allusion not only to William Morris and the bibliographical face of the late nineteenth-century aesthetic movement, but also to the longer tradition which those late nineteenth-century works were invoking: the tradition of the decorated manuscript and its Renaissance bibliographical inheritors' (1991, 122).

The optimal/ideal reader is a term used to refer to that collection of abilities, attitudes, experience, and knowledge which will allow a reader to extract the maximum value from a reading of a particular text. (For some commentators, the maximum *legitimate* value.) It should be noted that whereas for some critics the optimal/ideal reader is a universal figure, for most he or she is particular to given texts: the ideal reader for *Humphry Clinker* may not be the ideal reader for 'My Last Duchess'. Closely related again is the *informed reader*, a term given currency by Stanley Fish. According to Fish

> The informed reader is someone who (1) is a competent speaker of the language out of which the text is built up; (2) is in full possession of 'the semantic knowledge that a mature . . . listener brings to his task of comprehension,' including the knowledge (that is, the experience, both as a producer and comprehender) of lexical sets, collocation probabilities, idioms, professional and other dialects, and so on; and (3) has *literary* competence. That is, he is sufficiently experienced as a reader to have internalized the properties of local discourses, including everything from the most local of devices (figures of speech, and so on) to whole genres. (1980, 48)

In spite of the fact that all of these terms have sprouted on the grave (or the sick-bed) of the New Critics, it can be argued that all of them except the intended reader retain a certain text-centredness. Rather different is the *empirical reader*, used to describe those actual human beings who read a given literary work in varied ways and get varied things out of their readings. This concept arises perhaps more from the sociology of literature than from literary criticism as such, but study of the empirical reader and of empirical readings can have important implications for literary criticism. In particular, confirmation that one literary work can generate a range of different reading experiences, over time, between cultures or groups (or within them), and even for the same individual, leads necessarily to the question of the status and authority of these different reading experiences. If there is an optimal reader is there also an optimal reading, or is it a characteristic of major literature that it can generate a succession of new reading experiences as the individual reader or his or her culture changes? It has to be said that study of empirical readings is at a very early stage; Norman N. Holland's *5 Readers Reading* (1975) contained interesting material, but it could more accurately have been entitled *5 Readers Remembering What they Read.*

It is important to distinguish the reader or readers from the NARRATEE or NARRATEES, although in very special circumstances (a lyric poem addressed to and read only by a given person) the two rôles may be filled by one and the same person.

For *mediation reading* and *transparency reading* see OPAQUE AND TRANSPARENT CRITICISM.

See also COMPETENCE AND PERFORMANCE; NEW READERS; READERLY AND WRITERLY TEXTS; RECEPTION THEORY.

Reading as a woman / like a woman See READING POSITION

Reading community See INTERPRET[AT]IVE COMMUNITY

Reading position According to Anne Cranny-Francis, the concept of reading position helps us to understand how an audience is constructed by a literary TEXT. For her, a reading position is 'the position assumed by a reader from which the text seems to be coherent and intelligible' (1990, 25). Put another way: the text *situates* the READER in certain ways. Just as we have to get into a certain position to see certain physical objects, so too in order to read Jane Austen's *Pride and Prejudice* the reader has to adopt a stance with regard to the values and procedures of the NARRATIVE, unless a decision is made to give the WORK an OPPOSITIONAL READING.

Two related concepts are those of reading as a woman and reading like a woman. Evelyne Keitel distinguishes between the two as follows.

> 'Reading as a woman' rests on the (tacit) assumption that gender is the decisive factor in all human activity: women *are* different from men, therefore they *read* differently – reading is seen to be grounded in biology. Its second assumption is that there must be an obvious and *viable* continuity between all female experience . . .
>
> For women to read 'like a woman' requires an intentional and voluntary act of 'unlearning', an act of 'defamiliarization' with their gender roles.
>
> (1992, 371, 372)

Real See IMAGINARY / SYMBOLIC / REAL

Realism Debates around the concept of realism go back a long way, and recent arguments need to be seen as the latest attempts to grapple with certain perennial problems. (There is an excellent historical account of the development of this term – not limiting itself to literary or artistic usages – in Raymond Williams's *Keywords* [1976].) The most important elements in recent discussion of realism seem to stem from two sources: first the influence of what has become known as the Brecht–Lukács debate, and second what we can call the realist-imperative in FEMINIST theory.

The Brecht–Lukács debate dates from the 1930s, but Brecht's contributions were not made known to Lukács at that time nor were they published until much later. In his published writings the Hungarian MARXIST Georg Lukács had, by the mid-1930s, arrived at a definition of realism which placed a high premium upon the artist's (i) portraying the *totality* of reality in some form or other, and (ii) *penetrating beneath* the surface appearance of reality so as to be able to grasp the underlying laws of historical *change*. For Lukács, then, the artist's task is similar to that set for himself by Marx: to understand world history as a complex and dynamic totality through the uncovering of certain underlying laws. In practice this led Lukács to place a supreme value upon certain works of classical realism in the novel – the names of Tolstoy, Balzac and Thomas Mann are frequently on his lips or at the tip of his pen – and to wage an unceasing campaign against different aspects of what he saw as MODERNISM.

So that when Brecht, writing about the concept of realism, states that we 'must not derive realism as such from particular existing works' (Bloch *et al.* 1977, 81), the unstated target is clearly Lukács (especially as he goes on to refer to Balzac and Tolstoy). Brecht's argument is, in a nutshell, an anti-ESSENTIALIST one. In other words, realism for him is not intrinsic to a literary work, coded into it for all time like the name in a stick of rock, but a function of the rôle the work plays or can play in a given society at a particular historical moment. As Brecht says, reality changes and in order to represent reality modes of representation must also change (Bloch *et al.* 1977, 82). Brecht's realism focuses not on questions of form or content, but of function.

> Realistic means: discovering the causal complexes of society / unmasking the prevailing view of things as the view of those who are in power / writing from the standpoint of the class which offers the broadest solutions for the pressing difficulties in which human society is caught up / emphasizing the element of development / making possible the concrete, and making possible abstraction from it. (Bloch *et al.* 1977, 82)

Not surprisingly, then, Brecht's view of the formal experimentation associated with modernism is far more favourable than is Lukács's.

Interestingly, too, Brecht's views on this issue have something in common with those expressed by Roman Jakobson in his 1921 essay 'On Realism in Art'. Both stress the need for continuous formal regeneration and transformation in order to prevent AUTOMATIZATION, although it has to be said that Brecht's position has a practical-political element which Jakobson's lacks.

If the crudities of some theories of socialist realism inherited from the Soviet Union of the 1930s and 1940s led to a movement away from any concern with realism on the part of more recent theorists, the women's movement and associated feminist theories have placed it back on the agenda even when the term itself is not used. The essential point was made in an article by Cheri Register dating from the early years of contemporary feminist literary theory: 'Feminist criticism is ultimately cultural criticism' (1975, 10).

See also SYNTAGMATIC AND PARADIGMATIC.

Recall See ANALEPSIS

Receiver See SHANNON & WEAVER MODEL OF COMMUNICATION

Reception theory A term generally used in a relatively narrow sense to describe a particular group of (mainly German) theorists concerned with the way in which literary WORKS are 'received' by their READERS over time, but also sometimes used in a looser sense to describe any attempt to theorize the ways in which art-works are received, individually and collectively, by their 'consumers'. The names most frequently mentioned as core members of the particular group of theorists are Hans Robert Jauss, Wolfgang Iser, Karlheinz Stierle and Harald Weinrich.

The translators of an article by Karlheinz Stierle entitled 'The Reading of Fictional Texts' suggest that the German term *Rezeption* as used by those known as 'reception theorists', 'refers to the activity of reading, the construction of meaning, and the reader's response to what he is reading' (Stierle 1980, 83, n1). In analysing the activity of reading, reception theorists make considerable use of a concept defined by Hans Robert Jauss – along with Wolfgang Iser the most important founder-member of the group. In an early and highly influential essay entitled 'Literary History as a Challenge to Literary Theory', Jauss singles out three ways in which a writer can anticipate a reader's response, and these seem to represent his view of the component parts of the reader's *horizon of expectations*:

> first, by the familiar standards or the inherent poetry of the genre; second, by the implicit relationships to familiar works of the literary-historical context; and third, by the contrast between history and reality . . . . The third factor includes the possibility that the reader of a new work has to perceive it not only within the narrow horizon of his literary expectations but also within the wider horizon of his experience of life. (1974, 18)

All of these elements may fuse into a single if complex tradition which a TEXT accumulates as it becomes well-known, a tradition with which its reader has to contend or grapple as he or she reads and responds to it. A well-known text, in other words, raises more specific expectations for the reader than one that has accrued no such tradition. (The implication of the last-quoted sentence above is that the reader of an *old* work does *not* have to perceive it within the wider horizon of his or her experience of life, and this implication perhaps reveals a certain formalist element in the theory.)

An important part of the theory as it has developed and matured involves a view of literary texts as partially open, and of the responses engendered by them to be (again partly) the creation of their readers. Thus in a much-quoted example, Wolfgang Iser sees the sense a reader actively *makes* of a literary text to be contained within certain limits imposed by the text itself.

> In the same way, two people gazing at the night sky may both be looking at the same collection of stars, but one will see the image of a plough, and the other will make out a dipper. The 'stars' in a literary text are fixed; the lines that join them are variable. (1974, 282)

This argued element of textual containment has led the American critic Stanley Fish to distinguish his own concept of the INTERPRET[AT]IVE COMMUNITY from the reception theorists' horizon of expectations: for Fish the text is itself constituted by the pre-existing CONVENTIONS of the interpret[at]ive community.

Reception theory is distinguished by a certain interdisciplinarity, and its exponents make use of elements from aesthetics, Philosophy, Psychology, and (particularly), from PHENOMENOLOGY.

Reconstructionism The attempt to rebuild or expand the accepted CANON of literary works, normally by seeking to reduce or remove ethnocentric, SEXIST, or PRESENTIST bias or by countering such bias with oppositional bias(es).

Recuperation See INCORPORATION

Reductionism A pejorative description of any form of explanation which allegedly ignores higher-level qualities in whatever is being studied and reduces the object of study to certain or all of its lower-level qualities.

Redundancy Information theorists define redundancy as the degree of relative predictability of a message, as against the degree of unpredictability of a message, which is known as *entropy*. A message with a high degree of redundancy is one from which a large amount can be lost without the message being rendered incomprehensible.

It seems to be one characteristic of FORMULAIC literature that it is far more predictable, and of MODERNIST and POSTMODERNIST literature that it is far less predictable, than what one can posit as a literary NORM. Such variations in level of redundancy cannot be mechanically linked with aesthetic or other significance, however, although they may have relevance in discussion of different interpretations and of interpretation in general. On the other hand, the high redundancy level of the cliché is, except when it is used ironically, almost invariably given negative aesthetic value.

See also OPEN AND CLOSED TEXTS; READERLY AND WRITERLY TEXTS.

Reference / referent To refer is to point or allude to, and thus to assert the existence or nature of. In literary criticism from a very early time the term has been associated with controversies about whether literary WORKS make reference to extra-literary or extra-textual reality, and, if so, how. Many influential recent theories (and especially those associated with STRUCTURALISM and POSTSTRUCTURALISM) have argued that literature is non-referential, that statements in literary works cannot be either true or false because they refer to nothing that really exists – a position that can be traced back to Sir Philip Sidney's *An Apology for Poetry* (1595) and beyond.

J. Hillis Miller uses the term 'head referent' in a rather special sense, equating it with what DECONSTRUCTIONISTS call the (missing) CENTRE which would stop the play of SIGNIFIERS and which would allow the MEANING of the TEXT to be fixed if it existed (which they deny). Thus for Hillis Miller the head referent in Emily Brontë's *Wuthering Heights* is that 'original literal text' of which all the other texts that are in, and that constitute, *Wuthering Heights*, are figures. The reader of the novel searches for this head referent, according to Hillis Miller, in order to 'still the wandering movement from emblem to emblem, from generation to generation, from Catherine to Catherine, from Hareton to Hareton, from narrator to narrator' in the novel (1982, 67).

See also CENTRE; INTERTEXTUALITY.

Referential See FUNCTIONS OF LANGUAGE

Referential code See CODE

Reflector (character) Following Henry James, the central consciousness or intelligence in a work of fiction. More generally, a reflector is any consciousness or CHARACTER used by an AUTHOR to perceive things for the READER which otherwise would go unperceived.

See also PERSPECTIVE AND VOICE.

Refraction In the writings of Mikhail Bakhtin, according to the editor of *The Dialogic Imagination*, the intentions of the writer of prose are subject to refraction through 'already claimed territory' (Bakhtin 1981, 432). As the written word proceeds from writer to READER its direction is changed as a light-ray changes direction when it hits a prism. The metaphorical prisms in question exist both within the WORK (the other voices that the AUTHOR inhabits as the TEXT unfolds), and also outside the work: usages and associations in the reader's or critic's CULTURE through which the author's words have to struggle before making contact with the reader – in whose consciousness yet other refractions may take place.

Register The concept of register comes originally from the study of music, where it refers to the compass of a musical instrument or of a human voice. From here a succession of metaphorical adaptations take it to Phonetics where it is used to refer to the pitch of speech utterances, and to Linguistics in general in which it is used (with variations) to refer to context-dependent linguistic characteristics – either spoken or written, and encompassing any set of choices which are made according to a conscious or unconscious notion of appropriateness-to-context (vocabulary, syntax, grammar, sound, pitch, and so on). If, for example, one switches on a commercial radio station it is normally immediately apparent if an advertisement is being broadcast, for broadcast sound advertisements generally conform to particular register characteristics that make them immediately distinguishable as such, even if one hears them so badly that the actual words used are indistinguishable. There are, similarly, accepted (if changing) registers for church sermons, academic lectures, political speeches, declarations of undying love, and so on. An 'A-level' examination answer on Thomas Hardy's novel *Jude the Obscure*, marked by a friend of mine, which started 'What a miserable bugger Jude Fawley was', albeit undeniably accurate and dramatic, does not conform to the register to which such answers are normally expected to conform.

From Linguistics the concept has, by a further metaphorical leap, been adopted within literary criticism and NARRATOLOGY. Tzvetan Todorov, for example, in his *Introduction to Poetics*, lists a number of categories by which different *registers of DISCOURSE* may be recognized. These include the discourse's *concrete or abstract* nature, the absence or presence of *rhetorical figures*, the presence or absence of *reference to an anterior discourse* (giving, respectively, *polyvalent* and *monovalent* discourse), and, finally, the extent to which the language involved is characterized by 'subjectivity' or 'objectivity' (1981, 20–27).

It seems clear that the work done by the concept of register in Literary Studies to some extent duplicates that done by the concept of genre.

Reification According to the OED, the mental conversion of a person or abstract concept into a thing. In more recent usage the term has been applied to the process whereby relationships or processes are 'frozen' and treated as objects, and the usage resembles Marx's use of the term FETISHISM to describe the way in which a commodity is treated as a thing rather than as the visible point of a process of MEDIATION between groups or individuals.

In translations of Mikhail Bakhtin's work the term has been used to refer to that process of objectification which causes words to be treated as things with fixed meanings, rather than as mediators which gather new significance from use in varied contexts.

In literary-critical usage the term has been applied to the misconception, of which the New Critics were allegedly guilty, whereby the TEXT is fetishized at the expense of the shifting complex of relationships on which, it is argued, it depends.

Another very closely related term is *hypostatization*, meaning to reduce a process to a static substance; to turn a complex movement into a thing or object. The term thus has something in common with fetishism and reification, and the British literary theorist Christopher Caudwell tended to use it in his MARXIST writings in the 1930s in a way that was very close to Marx's use of the concept of fetishism.

Relation (maxim of) See SPEECH ACT THEORY

Relationism See ESSENTIALISM; RELATIVISM

Relativism The belief that there are no absolute standards, meanings, or truths, and that these have a force that is relative to their circumstances, contexts, or relationships. Generally speaking the term has a pejorative ring, and for this reason those wishing to defend a belief that all knowledge is of relations rather than of absolutes have preferred terms such as *relational* or the more modish DIFFERENCE.

Theoretical approaches such as STRUCTURALISM and DECONSTRUCTION have, many recent critics are agreed, met with problems when their essentially relational epistemologies have been extended from the realm of the abstract to that of more practical realms such as politics. Playing with TEXTS does not seem a morally adequate response to accounts of the holocaust.

See also ESSENTIALISM; HUMANISM.

Repertoire Often used in more specific ways in recent theory, as in 'cultural repertoires', 'symbolic repertoires', and so on. A person's CULTURAL repertoires would thus be composed of those cultural TEXTS, objects and practices to which he or she has expressive or participatory access. Thus were the Queen to enter a working-class pub she would probably find much that was outside her cultural repertoire.

Repetition The use of repetition is obviously central to a number of literary effects, and in consort with other elements such as metre, stress and rhythm in poetry it is a key means whereby the technical rate of REDUNDANCY is increased in a WORK. So far as poetry is concerned this can be related to ORALITY and the origins of poetry in oral performance: we repeat things more in speech than in writing to guard against mistakes. It would be a mistake to stop here, however, for repetition in literature has crucial aesthetic significance, and performs far more important functions than that of guarding against errors in transmission. The use of repetition can contribute to the formation of *patterns* in a literary work – THEMATIC, symbolic, and STRUCTURAL.

An influential study of the importance of repetition to fiction is J. Hillis Miller's *Fiction and Repetition*, which includes studies of seven central fiction texts from *Wuthering Heights* to *Between the Acts*. Hillis Miller distinguishes between two types of repetition, a distinction which he bases on Gilles

Deleuze's opposition of Nietzsche's concept of repetition to that of Plato. Deleuze summarizes the distinction in two formulations: 'only that which resembles itself differs', and 'only differences resemble one another' (Hillis Miller 1982, 5).

So far as NARRATIVE is concerned, Mike Bal reminds us that

> The phenomenon of *repetition*, . . . has always had a dubious side. Two events are never exactly the same. The first event of a series differs from the one that follows it, if only because it is the first and the other is not. Strictly speaking, the same goes for verbal repetition in a text: only one can be the first. . . . Obviously, it is the onlooker . . . who remembers the similarities between the events of a series and ignores the differences. (1985, 77)

Bal places repetition under the general heading of FREQUENCY, and the entry for this item should be consulted for information about types of repetition in NARRATIVE.

For the Freudian *repetition compulsion*, see the entry for FORT / DA.

Representatives See SPEECH ACT THEORY

Represented speech and thought See FREE INDIRECT DISCOURSE

Repression A key concept in Freudian theory, first used by Sigmund Freud in *Studies on Hysteria* which he wrote with Joseph Breuer. The concept is linked to that of censorship: repression involves the censorship of material such that the conscious mind becomes unaware of it, or is aware of it only indirectly (in jokes, dreams, 'slips', and so on). What is important is that the material is not lost, nor is it completely without effect upon the consciousness involved; its effect is indirect and unperceived.

The relevance of the concept to literature and literary criticism lies in the frequent suggestions by Freudians and neo-Freudians that art or literature have much in common with dreams and other activities which evade the censorship and release repressed material – either for the writer in the course of composition or for the READER during the reading process. A writer may not be able to recognize his or her own relief at the death of a parent, but may be able to express (and thus, to some extent, neutralize) the repressed material in disguised form in his or her writing. Such a theory clearly LEGITIMIZES (or hopes to legitimize) attempts to interpret a literary work by means of a psychoanalytic investigation of the AUTHOR.

Alternatively, a reader's responses to a literary work may be, according to some, traced back to his or her repressions. We would thus be able to express our emotions in uncensored form while reading a literary work. Analysis of our literary responses, then, would – like analysis of our dreams – help us to discover and identify our otherwise inaccessible repressions.

See also ARCHE-WRITING; DISAVOWAL.

Resonance A term used by the NEW HISTORICIST writer Stephen J. Greenblatt in preference, he explains, to 'allusion'– a term which he finds inadequate because it seems to imply a 'bloodless, bodiless thing' in contrast to the 'active charge' suggested by 'resonance' (1990, 163). What is being alluded to is the complex set or sequence of living meanings that a cultural artifact, or activity, (for example) can release when introduced into the right situation. We are close here to some of the issues discussed by Raymond Williams in connection with his use of the term STRUCTURES OF FEELING, and Greenblatt's acknowledged indebtedness to Williams may be relevant in discussion of his metaphorical use of the term 'resonance'.

According to Greenblatt, the New Historicism

> obviously has distinct affinities with resonance; that is, its concern with literary texts has been to recover as far as possible the historical circumstances of their original production and consumption and to analyze the relationship between these circumstances and our own. (1990, 170)

See also CIRCULATION.

Retroactive reading See MEANING AND SIGNIFICANCE

Retrospection / retroversion See ANALEPSIS

Revisionary ratios See REVISIONISM

Revisionism In current literary-critical usage this term is mainly associated with the theories of the American critic Harold Bloom, although it is still possible to find some more traditional MARXIST literary critics using it in the pejorative sense established in the writings of Lenin, where it refers to attempts to revise Marxism in such a way as to dilute or negate its revolutionary impetus.

In Harold Bloom's writings *revisionism* is used in a non-pejorative sense to indicate a general theory or set of theories concerning the manner in which the poet (Bloom uses this word in a wide sense) revises the work of his (Bloom's use of the female gender is very sparing) precursors. Thus, for Bloom, poetic influence 'is part of the larger phenomenon of intellectual revisionism' (1973, 28). Bloom sees this 'larger phenomenon' in terms of concepts much influenced by Freud; thus the poet's relation to these precursors is highly oedipal in character: the struggle against them is the struggle of the son against the father (or, to stress PATRIARCHAL authority, the Father). Just as Oedipus has to kill his father, the aspirant poet has somehow to destroy the power of his precursors, and, normally, of one especially potent patriarchal precursor, while simultaneously absorbing and transforming his strength and authority. The *strong* poet is the poet who succeeds in this task, and in his *The Anxiety of Influence* Bloom restricts himself to the strong poets, those who have the persistence to wrestle with their major precursors, 'even to the death' (1973,

5). Bloom's insistence upon the *anxiety* associated with influence is consistent with this view: the poet's attitude to his precursors is characterized by the same anxious mixture of love and rivalry to which the Freudian term *Oedipus complex* has been assigned. Bloom's use of Freud here thus shifts the literary-critical use of the oedipal theme from the level of literary content in the narrower sense to a more all-embracing biographical-content level. For Bloom, the poet suffers always from a sense of *belatedness*, a sense that he or she has come after important things have been said, and after they have been said in ways which constrain the poet and against which he or she has to struggle.

This leads Bloom to a rather different view of the process of READING. Restricting himself initially to the question of how the strong poet reads his precursor, Bloom asserts that poetic influence '*always proceeds by a mis-reading of the prior poet, an act of creative correction that is actually and necessarily a misinterpretation*' (1973, 30; Bloom's italics). But having argued for this position, Bloom is soon ready to extend it to the readings of the common READER and the literary critic, which he requires to be no less oedipal than those of the strong poet. He argues that most 'so-called "accurate"' interpretations of poetry are worse than mistakes, and suggests that 'perhaps there are only more or less creative or interesting mis-readings', because every reading is necessarily a *clinamen* – Bloom's term for a poetic misreading or misprision 'proper' (1973, 43).

Clinamen is the first of six 'revisionary ratios' outlined in *The Anxiety of Influence*. The others are as follows.

Tessera, or completion and antithesis, 'A poet antithetically "completes" his precursor, by so reading the parent-poem as to retain its terms but to mean them in another sense, as though the precursor had failed to go far enough'.

Kenosis, or a breaking device and movement towards discontinuity with the precursor, in which the poet's humbling of himself actually also empties out his precursor.

Daemonization, in which

> The later poet opens himself to what he believes to be a power in the parent-poem that does not belong to the parent proper, but to a range of being just beyond that precursor. He does this, in his poem, by so stationing its relation to the parent-poem as to generalize away the uniqueness of the earlier work.

Askesis, or a movement of self-purgation in which the poet yields up part of his own self-endowment so as to separate himself from others, including the precursor.

Apophrades, or the return of the dead, in which the poet holds his work so open to that of the precursor that the impression is given that the work of the precursor was actually written by the poet (1973, 14–16).

The extent to which these categories are available to use by critics other than Bloom is probably limited, but in more general terms *The Anxiety of Influence* has been Bloom's most influential work, and terms such as *revision-*

ism, *misprision*, and, especially, *misreading* have entered into current critical usage.

See also AGON; DOUBLE BIND; EPHEBE. For REVISIONIST HISTORY see next entry.

Revisionist history A term applied to a loose grouping of generally traditional historians who, to quote Richard Dutton, 'concentrate on such matters as aristocratic factions and the processes of patronage in propounding a view of Tudor and early Stuart politics (especially the politics of the court) very different from earlier accounts, both Whig and Marxist, which in their different ways had emphasised the corruption, imminent break-down and revolutionary potential of the system'. He notes that the revisionists 'have tended to stress the continuities and pragmatic accommodations within the system as it operated, providing a view of the practical mechanics of politics and power which in many ways is radically at odds with the New Historicist emphasis on ideologically charged structures of authority and subjection' (1992, 221–2). Dutton names as members of this grouping Conrad Russell, Kevin Sharpe, E. W. Ives, Linda Levy Peck, R. Malcolm Smuts, and David Starkey.

Rhizome See ROOT / RADICLE / RHIZOME

Root / radicle / rhizome Gilles Deleuze uses these three sorts of root growth to exemplify different types of connective relationship which epitomize different types of connective logic. The classical book is, according to him, the 'root book' which 'endlessly develops the law of the One that becomes two, then of the two that become four . . . Binary logic is the spiritual reality of the root-tree' (1993, 27, parenthesis in original).

The 'radicle-system, or fascicular root' is the figure of the book to which, he claims, our modernity pays willing allegiance (1993, 28). 'This time, the principal root has aborted, or its tip has been destroyed; an immediate, indefinite multiplicity of secondary roots grafts on to it and undergoes a flourishing development' (1993, 28). Deleuze cites William Burroughs's 'cut-up method' and the writing of James Joyce as productive of radicle-system books.

Finally, the 'rhizome' – a subterranean stem such as tubers have, offers a more complex model, one which, according to Deleuze, is also exemplified in the swarming rat-pack. Unlike trees or their roots, 'the rhizome connects any point to any other point, and its traits are not necessarily linked to traits of the same nature; it brings into play very different regimes of signs, and even nonsign states. The rhizome is reducible neither to the One nor to the multiple. . . . It is composed not of units but of dimensions, or rather directions in motion'. The examples Deleuze gives of rhizome structures are not books but, for example, the complex relationship between wasp and orchid ('a becoming wasp of the orchid and a becoming orchid of the wasp' [1993, 33]), and 'Benveniste and Todoro's current research on a type *C* virus, with its double connection to baboon DNA and the DNA of certain kinds of domestic cats' (1993, 36–7).

Russian Formalism Two related groups of theorists – the Moscow Linguistic Circle and the *Opojaz* group (= Society for the Study of Poetic Language) – are today generally seen to represent what is now seen as a single movement, and known as Russian Formalism. The title's connotations are now relatively neutral, but 'formalist' was used in a pejorative sense by the group's contemporaries, especially by those attacking the group from a MARXIST perspective. The most important Russian Formalists were Victor Shklovsky, Boris Eichenbaum, Boris Tomashevsky, Yuri Tynyanov, and Roman Jakobson – who was subsequently involved with the PRAGUE CIRCLE of theorists and eventually moved to the United States. (Different systems of transliterating Russian letters result in there being a number of different spellings of these names.) Vladimir Propp, author of *Morphology of the Folktale*, is not now generally classed as a member of the group, although he often was so classified at various times in the past, and Jakobson's work subsequent to his leaving the Soviet Union is, again, not normally classified as 'Russian Formalist' today.

The first sentence of Eichenbaum's essay 'The Theory of the "Formal Method"' gives a fair indication of its characteristic trajectory: 'The school of thought on the theory and history of literature known as the Formal method derived [Eichenbaum is writing in 1925] from efforts to secure autonomy and concreteness for the discipline of literary studies' (Èjxenbaum 1971a, 3). The autonomy in which the Russian Formalists are interested is less of the individual WORK and more of Literary Studies in general and of LITERARINESS; its CONCRETENESS is less what, following F. R. Leavis, we think of as the literary work's local effects and enactments of 'life', and more a matter of the isolation of technical devices and linguistic specificities. But it certainly shares with the New Critics a suspicion of literary criticism which relies upon biographical details about the AUTHOR or socio-cultural information about his or her age. To this extent it shares with the New Critics a desire to establish Literary Studies as a discipline independent from History, Philosophy, Biography, Psychology, and so on. But in its insistence upon the need for an independent *theory* of literature it stands at some distance from the New Critics and their view of what Literary Studies should be.

Three important formalistic contributions to Literary Studies need to be mentioned here: first, the belief in a distinction between poetic language and ordinary language – a distinction on which few today would bestow their agreement. Second, the importance of DEFAMILIARIZATION, a concept which has not yet outlived its usefulness. And third, the distinction between *fabula* and *sjužet* (see the entry for STORY AND PLOT), which plays a pivotal rôle in modern NARRATOLOGY.

One of the most potent critiques of the Formalists was actually first published in the Soviet Union in 1928: P. N. Medvedev's *The Formal Method in Literary Scholarship*, which some have claimed was actually written by

Mikhail Bakhtin (Medvedev was a member of Bakhtin's circle). Medvedev's book recognizes the positive elements in the work of the Formalists (his attack should not be equated with the Stalinist suppression of them), but he criticizes their rigid distinction between internal and external factors, and their inability to recognize that an external factor acting on literature can become 'an intrinsic factor of literature itself, a factor of its immanent development' (1978, 67). Medvedev is shrewd in his criticism of some Marxist attacks on the Formalists: it was not that the latter denied that external factors could influence literature, but that they denied that such factors could affect it intrinsically. Medvedev is also effective in his critique of the Formalists' commitment to the existence of a separate poetic language; as he puts it, 'The indices of the poetic do not belong to language and its elements, but only to poetic constructions' (1978, 86). However, as one would expect from a member of Bakhtin's circle, Medvedev's strongest criticism is reserved for the ahistoricism of Formalism, and for its failure to recognize that

> Even the inner utterance (interior speech) is social; it is oriented toward a possible audience, toward a possible answer, and it is only in the form of such an orientation that it is able to take shape and form. (1978, 97, 126)

Finally, he points out that although the Formalists attempted to escape from a psychological subjectivism in their approach to literature, their basic theories (including deautomatization) 'presuppose a perceiving, subjective consciousness' (1978, 149).

S

Salience See PUNCTUATION

Sapir-Whorf hypothesis The belief that any language will fundamentally and inevitably condition the way that those who have it as their mother tongue will see the social and material world. The name comes from the two American linguisticians, Edward Sapir and Benjamin Lee Whorf, and particularly from Whorf's studies of American Hopi Indians in the 1930s. Also known as linguistic relativism, the theory is an early example of the belief in – as the title of a book by Fredric Jameson has it – 'the prison-house of language'.

Satellite event See EVENT

Scene See DURATION

Scientism A form of REDUCTIONISM which examines all social and CULTURAL phenomena according to (normally pre-Einsteinian) scientific principles.

Script Monika Fludernik attributes this term to Frame Theory, instituted by Roger C. Schank and Robert P. Abelson in their book *Scripts, Plans, Goals and Understanding* (1977), and developed by Fludernik in a specifically NARRATOLOGICAL context.

The theory assumes that familiar events such as, for example, 'eating out' commonly involve a script, or a skeleton outline of what it is normal to report. Thus (unless there are exceptional reasons so to do), when reporting on a meal out one will typically assume that there was a waiter (and so not mention the fact), and will not normally give a description of the protagonists' eating processes (after Fludernik 1993, 447). Thus NARRATIVE accounts in fiction can often use particular scripts in deciding what information is redundant, or obvious, or whatever.

Scripts change: before Joyce, accounts of CHARACTERS' BODILY functions did not form a part of the scripts used by most serious writers.

Scriptible See READERLY AND WRITERLY TEXTS

Secondary process See PRIMARY PROCESS

Segmentation See PUNCTUATION

Self See SUBJECT

Self-consuming artifact From a book entitled *Self-consuming Artifacts* by Stanley Fish (1972). Literary TEXTS are self-consuming in the sense that affects produced in the course of a READING of them are destroyed as the reading proceeds. As Fish puts it in a later work,

> To read the *Phaedrus*, then, is to use it up; for the value of any point in it is that it gets *you* (not any sustained argument) to the next point, which is not so much a point (in logical-demonstrative terms) as a level of insight. It is thus a *self-consuming artifact*, a mimetic enactment in the reader's experience of the Platonic ladder in which each rung, as it is negotiated, is kicked away. (1980, 40)

Semantic axis A term coined by Mieke Bal in connection with her proposal for a method to determine which of a CHARACTER's characteristics are of NARRATIVE relevance and which are of only secondary importance. According to Bal, semantic axes are pairs of contrary meanings such as *large* and *small* or *rich* and *poor*. To determine which axes are productive of the most fruitful analytical results one has, Bal suggests, to concentrate upon 'those axes that determine the image of the largest possible number of characters, positively or negatively' (1985, 86).

Semantic position A term associated with translations of Mikhail Bakhtin's work, which seeks to convey the fact that a voice is not just a mark of specific human individuality but a commitment (recognized or unrecognized) to a given IDEOLOGICAL persuasion or persuasions. 'Position' here is of course a metaphor, and just as a physical position determines not just what we can see but also how we see what we can see, how we *situate ourselves in relation to the object of our view*, so too our adoption of a certain voice has (the argument runs) comparable implications on the ideological level. ('Voice' here must be given the wider meaning which Bakhtin habitually attributes to it: it is not just that people with, say, a middle-class accent tend to have middle-class views, but that the totality of a person's voice will reflect, imply and entail a set of ways of orienting oneself amongst other people in a particular context or contexts.)

A person's semantic position is supra-individual for another reason: it assumes and expects certain things of other people, and in turn places a certain pressure upon them either to conform to or to resist this placing.

Seme See SEMEME

Sememe A term coined by Umberto Eco to indicate a basic semiotic unit, by analogy with, for example, *phoneme* as the smallest intelligible unit of significant sound in a language. Thus a sememe would represent the smallest independent unit on the SEMIOTIC plane, although the significance of an individual sememe – as with the phoneme – can be refined, altered or disambiguated by processes of selection and combination.

Eco suggests, further, that '*a sememe is in itself an inchoative text whereas a text is an expanded sememe*' (1981, 18).

Eco's use of the term *styleme* represents a similar sort of coinage: a styleme is the smallest independent unit on the *stylistic* plane.

The term *seme* is to be found in the English translation of Roland Barthes's *S/Z*, where it is described by Barthes as being more or less interchangeable with the term *signifier* (see the entry for SIGN) on the semantic plane: 'semantically, the seme is the unit of the signifier' (1990, 17).

Compare IDEOLOGEME.

Semic code See CODE

Semiology / semiotics The term *semiotic* was coined at the close of the nineteenth century by the American philosopher Charles Sanders Peirce to describe a new field of study of which he was the founder, and *semiotics* traces its descent from this point. *Semiology* was coined by the Swiss linguistician Ferdinand de Saussure, and in his posthumously edited and published *Course in General Linguistics* he defended the coinage as necessary for the naming of that new science which would form part of social psychology and would study 'the life of signs within society' (1974, 16). (This placing of semiology in social psychology, incidentally, was subsequently to incur the wrath of Jacques

Derrida, who suggests that it is only by making the phonetic SIGN the pattern of all other signs [as he does], that Saussure can inscribe general semiology in a psychology – a step Derrida believes to be ill-conceived [1981b, 22].)

Today semiology and semiotics are generally used interchangeably, although attempts have been made to give each a distinct meaning. The latter term now appears to be rather more common, and I will thus adopt it in what follows. (Both terms are based upon adaptations of the Greek word *sēmeion*, meaning 'sign', and the spelling semeiology is sometimes encountered.)

Although Saussure was certainly no narrow formalist, semiotics has often been characterized by a formal-technical approach to the study of signs which, although it may have a place for terms such as *context*, has not emphasized social and CULTURAL determinants. On occasions many of Saussure's terms – SIGNIFIER and SIGNIFIED, LANGUE and PAROLE – have been recruited into the service of more or less acultural and asocial variants of semiotics, often under the aegis of STRUCTURALISM.

In contrast, other versions of semiotics have seen themselves, following Saussure's own suggestion, as a part of the Social Sciences rather than as an abstract, formal and technical discipline, and have seen their work to be closely related to Cultural Studies. It is significant, for example, that the English translation of Umberto Eco's essay 'Towards a Semiotic Inquiry into the Television Message' was first published in *Working Papers in Cultural Studies* in Britain, and that although this essay states a initial concern to consider television outputs as 'a system of signs', it also commits itself to a belief in the importance of empirical investigations into how viewers actually 'read' and understand what they see – what they actually 'get' (1972, 104).

For Jonathan Culler, semiotics is not primarily concerned to produce interpretations, but more to show how interpretations, or meanings, are generated.

> Semiotics, which defines itself as the science of signs, posits a zoological pursuit: the semiotician wants to discover what are the species of signs, how they differ from one another, how they function in their native habitat, how they interact with other species. Confronted with a plethora of texts that communicate various meanings to their readers, the analyst does not pursue a meaning; he seeks to identify signs and describe their functioning. (1981, vii–viii)

The work of Mikhail Bakhtin and his circle has renewed interest in the relationship between semiotics and IDEOLOGY. V. N. Vološinov's *Marxism and the Philosophy of Language*, for instance, which was first published in Russian in 1929, and appeared in English only in 1973, has a first chapter which is entitled 'The Study of Ideologies and Philosophy of Language', and it devotes particular attention to the nature of the *sign*.

> The domain of ideology coincides with the domain of signs. They equate with one another. Wherever a sign is present, ideology is present, too. *Everything ideological possesses semiotic value*. (1986, 10)

The work of Roland Barthes has been influential in both the more formalist and the more culturally oriented variants of semiotics during different stages of his intellectual career.

As suggested above, a use of the linguistic paradigm has often been crucial to semiotic analyses: widely different systems such as complete CULTURES (Lévi-Strauss), striptease (Roland Barthes), or the UNCONSCIOUS (Jacques Lacan) are examined as sign-systems operating like a language, and what a component of the system *is* is subordinated to, or explained in terms of, what semiotic function it performs. Not surprisingly, this can take semiotic analysis close to aspects of traditional literary analysis. Thus Robert Scholes's 'semiotic analysis' of 'Eveline', one of the stories in James Joyce's collection *Dubliners*, (Scholes 1982, 87–104) relies heavily on the codes of reading defined by Roland Barthes in his *S/Z* (see the entry for CODE).

Literary semiotics, like semiotics in general, comes in both formalist and cultural editions (to oversimplify somewhat). Robert Scholes points out that Yuri Lotman's *Analysis of the Poetic Text* (1976) and Michael Riffaterre's *Semiotics of Poetry* (1978), both 'approach poems through conventions and codes but share with the New Critics a sense of the poetic text as largely self-referential rather than oriented to a worldly context', whereas Barbara Herrnstein Smith's *Poetic Closure* (1968), although it also concentrates upon codes and CONVENTIONS as a way into the interpretation of texts, reveals her 'willingness to speak of a poem's "sense of truth"' that links her to other critics concerned with the emotional and intellectual impact of a text on READERS (1982, 12). An interest in the reader, then, is typical of what one can define as non-formalist literary semiotics. An interest in the AUTHOR is much less common among critics of this persuasion, many of them writing at a time when a belief in the death of the author was strongest.

Robert Scholes argues, thought-provokingly, that semiotics has become not so much the study of signs as the study of CODES, 'the systems that enable human beings to perceive certain events or entities *as* signs, bearing meaning' (1982, ix).

For Derrida's critique of Saussure's use of the term *semiology* see the entry for GRAMMATOLOGY.

Sender See SHANNON & WEAVER MODEL OF COMMUNICATION

Sense and reference The usual way in which a distinction made by the German philosopher Gottlob Frege between *Sinn* and *Bedeutung* is rendered in English, although sometimes *meaning and reference* has been preferred.

In the course of a complex discussion of the issue of what language can be said to refer to, Frege noted that phrases such as 'The Morning Star' and 'The Evening Star' can be said to share the same REFERENCE (that is, the planet Venus) but to be possessed of a different sense (or meaning). It is not just that the one phrase refers to the planet as seen in the morning, and the other to the planet as seen in the evening, but that each phrase has accrued a complex set of CULTURAL and literary associations of its own. Language, in other words, does not just divide the natural world up into segments which can then be referred to; it also encompasses human relations to that natural world along with a set of cultural and human MEANINGS which are in part created and in part reflected in language.

The distinction has appealed to aestheticians and literary theorists concerned to avoid two polarized responses to the question as to whether WORKS of literature or art can be said to refer to the real world (however defined). Such theorists reject the views that (i) the literary work makes no reference to the real world, but rather creates its own reality, and (ii) the literary work has no meaning independent of its reference to the real world. Instead, they argue that the literary work does have a reference to the real world, but that it also has a sense which is not wholly dependent upon or restricted to its reference. Another term often used to describe words with the same reference but a different sense is *partial synonymy* (see the discussion in Scott 1990, 108–14).

S'entendre parler A phrase associated with Jacques Derrida, which means literally 'to hear oneself speak'. Thomas Docherty (1987, 22) suggests that the following quotation from Walter J. Ong's *Orality and Literacy* (1982) is 'a more simple way of explaining the Derridean notion of *s'entendre parler* as fundamental to communication':

> Human communication, verbal and other, differs from the 'medium' model most basically in that it demands anticipated feedback in order to take place at all. In the medium model, the message is moved from sender-position to receiver-position. In real human communication, the sender has to be not only in the sender position but also in the receiver position before he or she can send anything. (Ong 1982, 176)

Ong explains that by the 'medium' model he means those approaches to communication which focus on 'media' and thus suggest 'that communication is a pipeline transfer of units from one place to another' (1982, 176).

For the medium model see FUNCTIONS OF LANGUAGE and SHANNON & WEAVER MODEL OF COMMUNICATION.

Sequence See FUNCTION

Sexism Maggie Humm defines sexism as 'a social relationship in which males denigrate females' (1989, 202–3), but this is arguably too restrictive a definition, for a number of reasons. First because in most current usage it is accepted that sexism has an existence on the IDEOLOGICAL plane as well as on the plane of actual social relationships, and second because current usage also often accepts the possibility that women can be guilty of sexism when they adopt PATRIARCHAL attitudes.

A broader definition would see sexism as a variant of ESSENTIALISM concerned with GENDER characteristics and usually relying upon STEREOTYPES, which attributes negative characteristics to women or females and positive characteristics to men or males. Humm's definition does have one very important virtue, however: it reminds us that sexism involves not just a form of insult but a means of repression. The ideology of sexism is thus linked to the practice of patriarchy. (In casual usage the term is often near-synonymous with *patriarchal*.) For this reason I would resist extending sexism to examples of essentialism which attribute negative characteristics to men or males: one might argue that these deserved to be labelled sexist, but in practice this is not representative of current usage so far as I can ascertain.

Many FEMINIST commentators single out Kate Millett's 1970 study *Sexual Politics* for special mention when discussing the way in which the term entered the vocabulary of just about all feminists and many others. Of interest to us is the fact that Millett focussed her argument upon literary TEXTS, and related the struggle for women's liberation to that sexism which could be uncovered in CANONICAL literary texts and criticism (her treatment of D. H. Lawrence was especially influential).

See also HOMOPHOBIA, which includes discussion of the term *heterosexism*, and HOMOSOCIAL.

Shadow dialogue See INTERIOR DIALOGUE

Shannon & Weaver model of communication

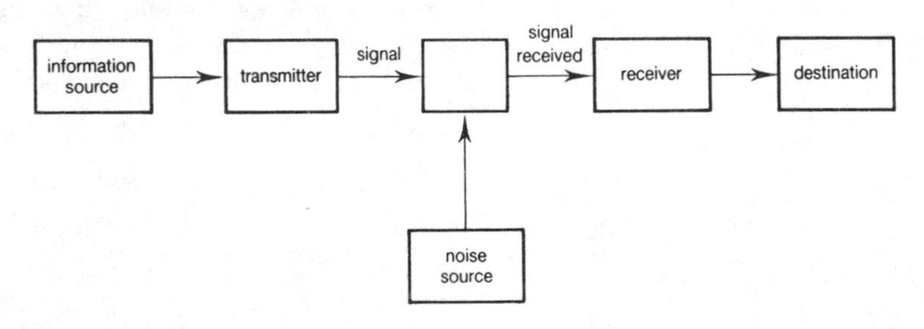

Shannon and Weaver model of communication, 1949

In 1948 the American electronic engineer Claude Shannon published two influential essays concerned to propose statistical ways of measuring the information-value of a message. Part of his argument included a diagrammatic model of the communication process, which has been extremely influential. As these essays were republished in book-form in 1949 with an additional essay by Warren Weaver, the model (see above) has become known as the Shannon and Weaver model.

Shannon and Weaver's model clearly aims at a high level of generality and abstraction, but it is important to remember that its creators were electronic engineers, used to working with information in a technical sense, information that could be *quantified* unproblematically. The model, however, seized the imagination of many working in very different fields, and the development of SEMIOTICS gave it a wider circulation than its creators had probably intended or anticipated. Its influence can almost certainly be detected in the influential passage in Roman Jakobson's 'Linguistics and Poetics' in which Jakobson attempts to break linguistic communication into its component parts: *addresser*, *message*, *addressee*, *context*, *code*, and *contact*. Jakobson's suggestion is discussed in the entry on FUNCTIONS OF LANGUAGE, in which the case against using such terms in the study of art or poetry is considered. It certainly seems possible to claim that the rôle of interpretation in the understanding of art makes this process rather different from the transmission of quantifiable information undertaken by the electronic engineer. The electronic engineer wants his or her 'message' to reach its 'destination' in unchanged form: the poet or novelist – even according to the most pro-intentionalist of theorists – does not want merely to recreate in the mind of the READER exactly what occurred in his or her mind during the composition of the work. It is probably not irrelevant to note that the word *message* has generally been used in a somewhat pejorative sense in discussions of art.

Perhaps a more successful adaptation of terms which probably originate with Shannon and Weaver can be found in recent NARRATIVE theory. Here a more schematic analysis of what in traditional terminology used to be called *point of view* (discussed in the entry for PERSPECTIVE AND VOICE) has proved to be genuinely illuminating, and in this context terms such as *addresser* and *addressee* serve a useful and generally non-REDUCTIVE function.

See also ACT / ACTOR.

Shifter In Linguistics: a linguistic unit which shifts REFERENCE according to the context in which it appears. Thus 'I', or 'the Pope' can both refer to a range of different historical or fictional individuals depending upon who utters them, to whom, and in what situation. The term has sometimes been used in discussions of literature, although not widely.

See also DEIXIS.

Short-circuit In NARRATIVE theory the deliberate and pointed mixing of narrative levels, or the flouting of CONVENTIONS appertaining to narrative levels. Much POSTMODERNIST fiction indulges frequently in such short-circuiting, as when an author speaks in his extra-fictional voice about one of his or her CHARACTERS.

Sign Much modern literary and critical theory has been dominated by or founded on a version of SEMIOTICS – that is, upon a general theory (or 'science') of the nature of the sign and of its life in CULTURE and history. Semiotics is not the child of literary criticism or theory, however, and no theory of the sign can be said to be literature-specific, or directed primarily or exclusively towards literature. Thus a concern with semiotic theories on the part of literary critics and theorists is both the result of a lessening belief in the specificity of literature and what can be clumsily called literary communication, and also the basis for a further search for what literary works and their reading have in common with what one must then refer to as 'non-literary communication'.

A useful starting point is to distinguish between *sign* and *symptom*. The key difference would appear to be that of the *conventional* nature of the sign: a symptom is fixed by and interpreted in the light of nature, a sign by and in the light of CONVENTION. Some theorists would be prepared to see a symptom as a sub-set of sign, while others would distinguish sharply between the two. It will be clear that the issue of intention, or MOTIVATION can be brought in here. (See the entry for SYMPTOMATIC READING in this context.)

Probably the theory of the sign which has been most influential so far as literature is concerned is that of the Swiss linguistician Ferdinand de Saussure. It is worth remembering that Saussure's definition is not of the sign as such, but of the *linguistic* sign, although many of his followers have generalized his definition to include non-linguistic signs as well. It should be remembered, however, that when one reads a statement along the lines of, 'Saussure defined the sign as . . .', one should bear in mind that this broadening of reference is neither unproblematic nor uncontroversial. (Saussure did, it is true, talk of the as then yet unborn science of semiology, which would study signs in general and their rôle in social life, but even here he closely associates the laws of semiology with laws applicable in Linguistics.)

Saussure denied the common-sense view that the linguistic sign was a name that could be attached to an object, arguing instead that the linguistic sign was a 'two-sided psychological entity' that could be represented by the diagram below (1974, 66; some translations have 'sound pattern' rather than 'sound image').

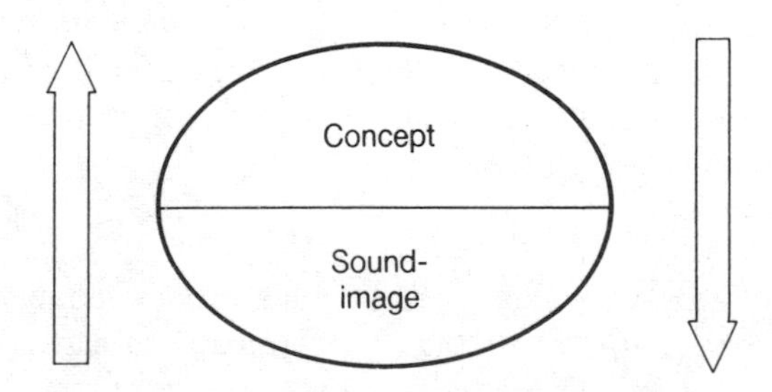

Saussure admits that this goes against then current usage (according to him the term *sign* was normally taken to designate only a sound-image), but argues that to avoid ambiguity three related terms were needed:

> I propose to retain the word *sign* [*signe*] to designate the whole and to replace *concept* and *sound-image* respectively by *signified* [*signifié*] and *signifier* [*signifiant*]. (1974, 67)

The English translations of *signifié* and *signifiant* given here have been questioned, and in one suggestion *significance* and *signal* have been proposed as alternatives.

It should be noted that subsequent theorists, Jacques Lacan included, have seen the link between signifier and signified to be far more problematic: shifting, multiple and context-dependent. And Jacques Derrida has seen Saussure's writing about the sign to double back on its own revolutionary insights; according to him, by equating the *signatum* (that which is signified) with the concept, Saussure leaves open the possibility 'of thinking a *concept signified in and of itself*, a concept simply present for thought, independent of a relationship to language, that is of a relationship to a system of signifiers' (1981b, 19). This, Derrida claims, allows Saussure to fall back into the 'classical exigency' of the TRANSCENDENTAL SIGNIFIED (1981b, 19). In other words, the concept is seen to be possessed of an identity separate from the defining system of differences between signifiers: it is seen as extra-systemic and complete in itself. (See the entry for RADICAL ALTERITY.)

It will be noted that what drops out in the movement from what I called the common-sense view to Saussure's definition is the object, or that which represents extra-linguistic reality. This has led many Saussureans to claim that language has no connection with extra-linguistic reality, an assertion that has been advanced especially forcefully by some *literary* critics and theorists concerned to adapt Saussure's theory of the sign to literature. Such assertions have very often brought in Saussure's argument concerning the ARBITRARINESS of the sign in support of this thesis, as well as the claim that Saussure's Linguistics is purely SYNCHRONIC and that it rejects the validity of any DIACHRONIC or historical, study of language. I have argued elsewhere (Hawthorn 1987, 52–7) that all of these assertions are incorrect, and that Saussure specifically rejects the basis for all of them in his *Course in General Linguistics*. The acceptance of all or some of these assertions as correct, however, has often provided a theoretical basis for a new formalism since the 1960s, a formalism that involves the isolation of literature from life, and art from society, culture and history. One quotation will perhaps suffice to confirm this; writing about Saussure's theory of the sign in his book *Semiotics and Interpretation* Robert Scholes states confidently that

> Saussure, as amplified by Roland Barthes and others, has taught us to recognize an unbridgeable gap between words and things, signs and referents. The whole notion of 'sign and referent' has been rejected by the French structuralists and their followers as too materialistic and simpleminded. Signs do not refer to things, they signify concepts, and concepts are aspects of thought, not of reality.
> (1982, 24)

To do Scholes justice it should be said that he goes on to challenge this 'recognition', although he does not challenge the justice of blaming it on Saussure. His summary is otherwise a fair account of an orthodoxy which flourished in the 1970s and 1980s.

A good summary of other important theories of the sign is given by Edmund Leach in the second chapter of his book *Culture and Communication* (1976, 9–16). Leach cites not just Saussure, but also C. S. Peirce, Ernst Cassirer, L. Hjelmslev, Charles Morris, Roman Jakobson and Roland Barthes as influential theorists in this context, but allots the word 'sign' only a limited function in his own terminology, which he represents by means of a complicated grid diagram. (Roland Barthes is driven to the same resort, which I have myself deliberately avoided, in his early work *Elements of Semiology*.) Jonathan Culler has pointed out (1981, 23) that Peirce's influential tripartite distinction between icon, index and symbol is only one of the many semiotic taxonomies he proposed, and that he finally committed himself to a core 66 classes of sign – a figure which has not generally been adopted by subsequent theorists.

In his later work Roland Barthes has suggested more idiosyncratic definitions for related terms: in *The Pleasure of the Text*, for example, he defines *significance* [*signifiance*] as 'meaning, *insofar as it is sensually produced*' (1976, 61) – a definition that, not surprisingly, has not achieved widespread popularity. There is a useful discussion of the use by both Barthes and Julia Kristeva of the French terms *signifiance* and *signification* in Richard Harland's book *Superstructuralism*; Harland suggests that what characterizes this usage is a restriction of the latter term to a fixity and self-identification within a system, and the opening out of this fixity of meaning to a decomposed and more INTERTEXTUAL *signifiance*. Again, such specialist usages have enjoyed only a limited influence in the work of other, non-French writers.

Gérard Genette has suggested an adaptation of Saussure's definition of the sign to the needs of the theory of NARRATIVE; he proposes 'to use the word *story* for the signified or narrative content . . . to use the word *narrative* for the signifier, statement, discourse or narrative text itself, and to use the word *narrating* for the producing narrative action (1980, 27). The vocabulary proposed is useful, but whether it depends upon the analogy with Saussure's definition of the linguistic sign is debatable.

Jacques Derrida has argued that not a single signified escapes 'the play of signifying references that constitute language'.

> The advent of writing is the advent of this play; today such a play is coming into its own, effacing the limit starting from which one had thought to regulate the circulation of signs, drawing along with it all the reassuring signifieds, reducing all the strongholds, all the out-of-bounds shelters that watched over the field of language. This, strictly speaking, amounts to destroying the concept of 'sign' and its entire logic. (1976, 7)

Others, however, have felt that this obituary is premature.

See also SEMEME.

Signifiance / signifiant See SIGN

Significance See SIGN; MEANING AND SIGNIFICANCE

Signification / *signifié* / signified / signifier See SIGN

Signifying practice In his Introduction to Julia Kristeva's *Desire in Language* Leon S. Roudiez quotes Kristeva's definition of this term from her *La Traversée des Signes*:

> I shall call signifying practice the establishment and the countervailing of a sign system. Establishing a sign system calls for the identity of a speaking subject within a social framework, which he recognizes as a basis for that identity. Countervailing the sign system is done by having the subject undergo an unsettling, questionable process; this indirectly challenges the social framework with which he had previously identified, and it thus coincides with times of abrupt changes, renewal, or revolution in society. (Kristeva 1980, 18)

For Roland Barthes, a signifying practice is 'first of all a differentiated signifying system, dependent on a typology of significations (and not on a universal matrix of the sign)'. This means, he continues, that signification is not produced at the abstract level of LANGUE, but through 'a labour in which both the debate of the subject and the Other, and the social context, are invested in the same movement' (1981b, 36). According to Barthes, this notion restores its active energy to language.

A signifying practice thus involves a struggle for MEANINGS within a social context and against other interests, a struggle which is inseparable from the identity granted or claimed for him or herself by the individual and others.

See also DISCOURSE; IDEOLOGY; UTTERANCE.

Singularization See DEFAMILIARIZATION

Singulative See FREQUENCY

Sinn und Bedeutung See SENSE AND REFERENCE

Site A word that has accrued a particular polemical force in recent years, although a force that, paradoxically, is linked to a claimed neutrality, IDEOLOGICAL innocence, and lack of proper energy or initiatory force. To describe something as a site has become a favoured way of giving (sometimes exclusive) precedence to external determining forces while playing down or denying the self-initiation of movement or development.

Thus when the SUBJECT is described as site, an attack is being made on views of the subject as in control – or even aware – of its own destiny; a subject is, according to such a view, more or merely a site on which extra-subjective forces clash and resolve their differences. These extra-subjective forces which are given priority can be either language, or history, or ideology, or the class struggle, or whatever. Put simply, the argument thus invoked by the use of *site* denies that the individual is in charge of his or her life, or consciousness, and asserts rather that the subject is constituted by other forces which then proceed either to nest or to fight in it.

Those fond of the term have sometimes come dangerously near to a sort of infinite regress: the subject is a site upon which different ideological forces battle, but then ideology is itself a site on which different class interests war with one another, while class is, in its turn, a site on which different relationships to the forces of production meet in conflict, and so on.

The term is sometimes invoked by those DECONSTRUCTIONISTS arguing for a rejection of any TRANSCENDENTAL SUBJECT or PRESENCE.

In literary criticism the term is often associated with attempts to downgrade the extent to which the AUTHOR is seen to be in control (conscious or otherwise) of his or her creation.

Sjužet See STORY AND PLOT

Skaz A mode or technique of narration that mirrors oral NARRATIVE. A useful alternative definition of the term is given by Ann Banfield, who points out that apart from the epistolary form, *skaz* 'is the only type of literary first person narrative which clearly has a second person' (1982, 172). The term comes from the Russian, as RUSSIAN FORMALIST theorists were the first to tackle the issues raised by *skaz*.

For the Russian Formalist Boris Eichenbaum, *skaz* allots – or can allot – a special rôle to aspects of a NARRATOR's verbal delivery such as articulation, miming, and sound gestures. Eichenbaum also distinguishes between the concept of PLOT and the concept of STORY-stuff, or, alternatively, between construction and material. *Skaz* for Eichenbaum is important because it is 'the constructional principle of the plotless story' (Èjxenbaum 1971a, 21). (See the entry for STORY AND PLOT.) Mikhail Bakhtin takes issue with Eichenbaum, claiming that the most important thing about *skaz* is not '*orientation toward the oral form of narration*', but 'orientation toward *someone else's speech*' (1984, 191).

Ann Banfield adds a number of other, more complex points. She claims that *skaz* is not actually a form of narration, but 'the imitation of a discourse'. For her, DISCOURSE is characterized by the inseparability of telling from expression, whereas the function of a sentence in narration is 'solely to tell' (1982, 178). Her claim is by no means uncontroversial, and relies upon definitions of *discourse* and *narration* that would not find universal acceptance. She makes one additional, important technical claim: represented speech and thought (see the entry for FREE INDIRECT DISCOURSE) are not found in oral literature, or literature derived from the oral – and this includes *skaz* (1982, 239–40).

Slant See PERSPECTIVE AND VOICE

Slippage The term seems to be an anglicizing of the French *glissement*. Vincent Crapanzano suggests, if I read a complex passage aright, that in the work of Jacques Lacan it is applied to that distortion that takes place between the dream and the account given of it by the (woken) dreamer, a distortion which is 'a sliding, a *glissement* of the signified under the signifier, the concept under its acoustic image' (1992, 145).

More loosely: the (normally unconscious) redefinition of terms or commitments in the course of an argument, often as a result of IDEOLOGICAL pressure.

Slow-down The narrative equivalent to that slow-motion experience which frequently accompanies moments of great danger or tension. A narrative may suddenly provide what seems (in contrast to that which has gone before or that which comes after) a disproportionate number of pages to what takes place in a (relatively) very short span of time. Precisely because the technique does resemble 'slow-motion experience' it suggests by analogy that what is narrated is of great subjective importance either to narrator or CHARACTER, or both.

Social energy See ENERGY

Socialist realism See REALISM

Sociolect Formed by analogy with *dialect* (compare IDIOLECT and GENDERLECT), and used to denote language (PAROLE rather than LANGUE) which is specific to a particular social group, and which thus carries with it the values and status of the same group. The social group in question can be defined in terms of class (variously understood), age, or GENDER – or a permutation of all three. Sociolects are normally associated with speech, although as lexical, grammatical and syntactical elements can serve to isolate a sociolect it is possible to trace sociolects in written TEXTS.

Socio-linguistic horizon See HORIZON

Solution from above / below Also known as 'hypothesis driven' (above) and 'data driven' (below) solutions, and also 'top down (perspective)' and 'bottom up (perspective). From the psychology of visual perception: a solution from above is a perceptual interpretation which has been arrived at as a result of a hypothesis which predates the input of sensory data; a solution from below is a perceptual interpretation which has been prompted by the actual input of sensory data which is itself then interpreted.

Solution from below research starts, as the term 'data driven' suggests, from some form of evidence, from what, perhaps misleadingly, can be called 'raw material' – normally raw material that requires some sort of explanation or ordering. In this sort of research, theory is *brought in as part of the solution.* When a couple goes to marriage counselling, for example, they are engaged in data driven research. Hypothesis driven research, in contrast, takes place independently of or prior to direct contact with the 'raw material' which one wishes to study; it stems from a problem within theory, from pressures at a more abstract and displaced level than the problems that initiate data driven research. In this sort of research it is *the raw material which is brought in as part of the solution.*

A classic example of such research would be the predictions about planetary movement entailed by Einstein's theories which were eventually confirmed by directed observation. Classically, the pure sciences work in terms of hypothesis driven research, while the applied sciences work in terms of data driven research.

By extension, a literary interpretation which is 'from above' or hypothesis driven is one prompted by a hypothesis which the WORK itself, or the readings of 'Common readers', do not overtly suggest; a literary interpretation which is 'from below' or data driven is one which is prompted by the work itself.

The distinction seems more straightforward than it is, as what a work 'overtly suggests' is arguably dependent upon the READER's expectations, critical awareness, social and CULTURAL background, and so on. Nevertheless, the distinction has its merits, especially during the present time when many critical theories propose the need for the critic to step outside a *naturalized* understanding of, or RESPONSE to, the work of literature, and to give it instead an OPPOSITIONAL READING.

In this context it is worth noting that recent literary theory has often obscured the distinction between *empiricism* (the belief that direct experience or observation is the only source of knowledge), and *empirical* research (which relies upon such direct experience or observation but does not necessarily rest upon the view that such experience or observation is the *only* source of knowledge, or that its significance is self-evident without the application of non-empirical ideas or theories.

Sous rature See ERASURE

Spacing See PUNCTUATION

Speech Attempts to render Ferdinand de Saussure's distinction between LANGUE AND PAROLE into English frequently use *speech* as the preferred term for *parole*. However, it is increasingly common to use the original terms in untranslated form, in part because *speech* is a rather unsatisfactory translation of *parole* – something like *the sum of all actual (or possible)* UTTERANCES would probably come nearer to what Saussure seems to have had in mind. An utterance, like *parole*, does not need to be actually spoken: it can be written or expressed to oneself in silent thought.

We can compare the problems associated with translating ÉCRITURE as *writing*, discussed in the entry for this term.

As a general principle, then, it is worth exercising care when one encounters words such as *speech* and *writing* in English translations of French theoretical writing. Jacques Derrida's commitment to the primacy of Writing over Speech, for example (reiterated in a number of his texts) seems less bizarre when his original terms are borne in mind: he is certainly not saying that human speech is historically subsequent to writing in the everyday meanings of these words.

See also SPEECH ACT THEORY; VOICE.

Speech act theory As its name suggests, speech act theory is a theory which attempts to explain exactly what happens when human beings speak to one another. The theory has been influential not just within Linguistics and PRAGMATICS, but also within Literary Criticism – as a result of books such as Mary Louise Pratt's *Toward a Speech Act Theory of Literary Discourse* (1977). Speech act theory originated with the philosopher John Austin's book *How to do Things with Words* (1962), in which Austin argues against the philosophical assumptions that verbal statements can be analysed in isolation and in terms only of their truth or falsity.

A number of other philosophers – most notably John Searle but also H. P. Grice and P. F. Strawson – have developed and extended Austin's arguments. They have drawn attention to the manner in which the public uttering of statements is governed by rules and CONVENTIONS which have to be understood and abided by on the part of utterers and listeners if effective communication is to take place, and also to the fact that statements not only *are* things, but they typically also *do* things. In Searle (1969) we find a useful distinction between a number of different sorts of verbal act:

> *(a)* Uttering words (morphemes, sentences) = performing *utterance acts*.
> *(b)* Referring and predicating = performing *propositional acts*.
> *(c)* Stating, questioning, commanding, promising, etc. = performing *illocutionary acts*. (1969, 24)

Searle's usage is not identical to Austin's: although Searle uses Austin's term 'illocutionary act', for example, he does not accept Austin's distinction between locutionary and illocutionary acts (1969, 23 n1).

Austin distinguishes between *constatives* – utterances which can either be true or false because they claim to report that certain things are the case in certain worlds – and *performatives*, which are utterances used to do something rather than to say that something is the case (e.g. 'I promise to marry you') and of which it makes no sense to claim that they are either true or false. One should add that a constative can also have a performative aspect – by pragmatic implication, for example.

Utterance acts are also referred to as *locutionary acts* if they involve the production of a recognizably grammatical utterance within a language community. *Illocutionary acts* include such things as asserting, warning, promising and so on, and Austin claimed that over a thousand different possible such acts can be performed in English. Searle has suggested that illocutionary acts can be classified into five basic categories: *representatives*, which undertake to represent a state of affairs; *directives*, which have the aim of getting the person addressed to do something; *commissives*, which commit the speaker to doing something; *expressives*, which provide information about the speaker's psychological state, and *declarations*, which bring about the state of affairs they refer to (e.g. 'I now declare you man and wife') (1976, 10–14).

Searle has also pointed out that a person who performs an illocutionary act may also be performing what he has dubbed a *perlocutionary act*, in other words, achieving certain intended results in his or her listener. He gives as examples, *get him to do something, convince him (enlighten, edify, inspire him, get him to realize)* (1969, 25).

The conventions underlying successful conversation are known as *appropriateness conditions* or *felicity conditions* by speech act philosophers, and together are taken to constitute the *co-operative principle* (a term suggested by H. P. Grice) that governs conversation in ideal situations. They include (i) *maxim of quantity*: (Make your contribution no more or less informative than is required); (ii) *maxim of quality*: (Make your contribution one that is true, and that does not include either false or inadequately substantiated material); (iii) *maxim of relation*: (Be relevant); (iv) *maxim of manner*: (Be perspicuous, avoiding obscurity, ambiguity, unnecessary prolixity). (Based on Pratt 1977, 130, following Grice.)

On the basis of such maxims, and on their assumed observance by participants in a conversation, statements not immediately interpretable as relevant can be seen to have a particular *implicature* or implication. Take the following exchange between husband and wife: 'Do you want to go home?' 'It's getting rather late'. The second statement may seem to be changing the subject, but its pragmatic meaning or implicature is clearly 'Yes'.

A theory of utterances which reduced the importance of truth REFERENCE and emphasized instead the compact formed between utterer and listener by the mutual adoption of a set of conventions, made literary critics sit up and take notice, for it seemed of far greater potential use in literary criticism than more traditional philosophical approaches to verbal statements had been. An obvious application of such a theory was to the conversations between CHARACTERS in plays and literary NARRATIVES, but there were more interesting and sophisticated ones. In his essay 'The Death of the Author' Roland Barthes suggested that the word writing (see ÉCRITURE) no longer designated 'an operation of recording, notation, representation, "depiction"', but rather 'a performative . . .

in which the enunciation has no other content . . . than the act by which it is uttered' (1977, 145–6; see also the entry for AUTHOR).

In a very different way the theory of implicature seemed especially promising for literary-critical application. When a writer wrote something that seemed to be irrelevant to what the READER had been led to be interested in, was this not because the writer could rely upon the reader's *searching* for relevance in the TEXT? Do not readers assume that everything that is there in a literary text is there for some purpose, has in other words some implicature?

Some critics have suggested that speech act theory tends to stress co-operation at the expense of struggle and disagreement. Most conversations are not conducted between equals whose interests are identical; they are conducted between individuals who have divergent interests and who either possess, or are subject to the exercise of, social power and authority. Most participants in conversations thus, the argument continues, typically break many of the maxims quoted above to further their own interests, and they expect their fellow conversationalists to do the same. Much the same is true of literary reading: the author and the reader have different interests and are as much trying to outwit as to co-operate with one another. In her poem 'Murder in the Dark' Margaret Atwood compares the literary process to the party game referred to in her title, with the writer as murderer and the reader as detective.

Speech act theory has probably had a positive effect upon literary critics inasmuch as it has stressed the importance of mutually accepted conventions in the reading and writing of literature, and it has refocussed attention on to the conventional elements implicit in LITERARINESS.

Speed See DURATION

Spots of indeterminacy See CONCRETIZATION

Staircasing See EVENT; FRAME

Stereotype Originally taken from a process in printing, the term has become a Standard English phrase for something that is fixed and unchanging – normally with a pejorative ring. It has played an important part in recent FEMINIST theory in connection with the description of fixed and, normally, PATRIARCHAL or SEXIST views of GENDER rôles and characteristics.

Thus the central third chapter in Mary Ellmann's influential book *Thinking About Women* (first published 1968) is devoted to 'Feminine Stereotypes', and lists formlessness, passivity, instability, confinement, piety, materiality, spirituality, irrationality and compliancy – along with 'two incorrigible figures': the shrew and the witch. Ellmann's list has been discussed and amended by subsequent writers, and it is fair to say that an investigation into stereotyping plays a key rôle in the discussion of a range of IDEOLOGICAL processes.

Stereotyping is not directed only at women, of course: we can isolate stereotypes based on race or CULTURE, on age, on profession, and so on.

Moreover, stereotypes may masquerade as positive: thus the beliefs that women are naturally intuitive, or that Black people are always happy and have a wonderful sense of rhythm, are no less stereotypical than more overtly negative views. (It will be noticed that such 'positive' stereotypes typically serve to advance more concealed, negative portrayals.)

Stereotypes are, it has been argued, part of the means whereby SUBJECTS are INTERPELLATED by ideological forces.

Story See STORY AND PLOT

Story and plot With this item we are concerned with an essentially very simple distinction surrounded by minefields of confusing vocabulary. The simple distinction is between, on the one hand, a *series of real or fictitious events, connected by a certain logic or chronology, and involving certain ACTORS*, and on the other hand, the *NARRATION of this series of events*. Thus were one to be asked to give the story of *Wuthering Heights* a suitable response would be to start with the first arrival of the child Heathcliff at Wuthering Heights and then proceed to recount the events of the novel in chronological order until the death of Heathcliff and the (possible) reuniting of him with Cathy. But the plot of *Wuthering Heights* is these events *as they are actually presented in Emily Brontë's novel*. Clearly the same story can give rise to many different plots, as FORMULAIC literature reveals very clearly.

The minefields of which I spoke arise from the fact that the same distinction is also referred to by means of the Russian words *fabula* (story) and *sjužet* (plot), and in some translations these are rendered in ways that conflict with the usage associated with story and plot. Mieke Bal, for example, renders our story and plot as *fabula* and *story* (1985, 5), which gives *story* the opposite meaning from that with which we started. Meanwhile other translators of RUSSIAN FORMALIST texts have suggested that fabula be rendered as *plot* – thus further giving also this term exactly the opposite meaning from the plot in our pair plot and story. (Other translators have suggested *fable and subject* for fabula and sjužet, which introduces yet further possibilities of confusion.)

Speaking as one who has himself trodden on some of these mines, my advice is as follows. When reading these terms always proceed with care and try to confirm what CONVENTION of usage the writer is following. When writing, explain your own convention of usage by making reference to what seems to be the one reliably unambiguous term: the paired *fabula and sjužet*. Do not use terms such as *story* or *plot* on their own without making it very clear to what you are referring.

A good illustration of the complexities of terminology that are here involved can be found in Monika Fludernik's diagrammatic representation of the terminology utilized by different narrative theorists to represent what she refers to as the different *narrative levels*:

	Events in chronological order	Events causally connected	Events ordered artistically	Text on page	Narration as enunciation
Genette	*histoire*	*histoire*	*discours (récit)*	*discours (récit)*	narration (voice + focalization)
Chatman	story	discourse	discourse	discourse	discourse
Bal	*fabula*	story and focalization		narration (+ language, + voice)	
Rimmon-Kenan	story	story	text	text	text
Prince	narrated	narrated	narrated	narrating	narrating
Stanzel	–	story	story	mediation by teller or reflector + enunciation if teller figure	

(Based on Fludernik 1993, 62)

I cannot resist closing this entry with a comment by the novelist Ivy Compton-Burnett, who I am glad to say follows the usage I recommend in the first paragraph above:

> As regards plots I find life no help at all. Real life seems to have no plots at all, and as I think a plot desirable and almost necessary, I have this extra grudge against life. (Quoted in Furbank 1970, 124)

String of pearls narrative A NARRATIVE that consists of a number of relatively or completely unrelated episodes strung together by a thin thread. The thread can consist of a causal sequence, or the person of an individual CHARACTER, or whatever. Otherwise known as an *episodic* narrative.

Strong poet See REVISIONISM

Strong textualist See TEXTUALIST

Structuralism It is worth starting off by distinguishing between modern structuralism, which is essentially a post-Second World War development, and the structuralism of the PRAGUE SCHOOL theorists (for which see the entry for Prague School). Although these have something in common – notably a central and substantial debt to the work of Ferdinand de Saussure – their differences are such as to justify our treating them separately. Modern structuralism, in its initial phase in the 1950s and 1960s, is a largely French phenomenon whose most important figures are probably the anthropologist Claude Lévi-Strauss, and the literary and cultural critic Roland Barthes. Following them, however, many other individuals produced work which has been labelled structuralist. These include a group of MARXIST theorists of differing persuasions within Marxism such as Louis Althusser and Lucien Goldmann (originally Romanian and known as a 'genetic structuralist'); NARRATOLOGISTS such as Gérard Genette, and Michel Foucault who, for want of a better classification, we can call a historian. We can add that in his insistence that the UNCONSCIOUS is structured like a language, Jacques Lacan also advances a classic structuralist position.

Common to all of these is an interest in structures or SYSTEMS which can be studied SYNCHRONICALLY rather than in terms of their emergence and development through traceable processes of historical causation (this applies even to Foucault, but only with qualifications to Goldmann), and a debt to the theories of Saussure – often involving a commitment to the implications of the LINGUISTIC PARADIGM. Put crudely, structuralism is (at least in its early or 'pure' form) interested rather in that which makes MEANING possible than in meaning itself: even more crudely – in form rather than content.

Not surprisingly, then, many structuralist ideas can be traced back to the RUSSIAN FORMALISTS, either directly or as mediated through the Prague School. A classic early example of a structuralist involvement with literature (or at least with NARRATIVE, depending upon how one defines literature) is to be found in Vladimir Propp's *Morphology of the Folktale*, first published in Russian in 1928. Typical features of Propp's study are its concern to generalize features (or FUNCTIONS) across TEXTS, and thus to concentrate upon a system capable of generating meanings which goes beyond the confines of the individual WORK, and its concomitant lack of interest in the interpretation of individual works or, even, in their individual specificity. (Roland Barthes's proposed CODES OF READING can be seen to be the direct descendants of Propp's functions.) From this perspective it is as if the work is written not by its AUTHOR (collective or individual), but by the 'grammar' or system of transformations that pre-exists its creation. A fundamental distinctive feature of structuralism can perhaps be indicated by comparing this to the way in which Lévi-Strauss can treat the system of gift-giving in a CULTURE, or Roland Barthes can treat 'Steak and Chips' or 'Striptease' in his *Mythologies*. What all of these studies have in common is a concern with the pre-existing system which allows individual UTTERANCES to be made. Indeed, use of the linguistic paradigm means that for the structuralist a particular meal, or giving of a gift, can be treated as a sort of utterance, an example of PAROLE behind which lies a complete LANGUE. When we give a visitor a meal of steak and chips, the meaning which this has for him or her is determined not just by (or not perhaps even primarily by) the actual taste and appearance of the meal, but the grammatical function which 'steak and chips' is allowed to play in the *langue* of meals. How and why this *langue* has developed into the system that exists is a matter of less (or no) interest to structuralism.

Clearly, this downplays the importance of 'the meal itself' and of the individual cook's skill, and thus it is not surprising that when structuralism is applied to literary works, it downgrades or rejects the view that either the 'work itself' or the author determines how the work is to be read. As Genette puts it,

> The project [of Structuralism], as described in Barthes's *Critique et verité* and Todorov's 'Poétique' (in *Qu'est-ce que le structuralism?*), was to develop a poetics which would stand to literature as linguistics stands to language and which therefore would not seek to explain what individual works mean but would attempt to make explicit the system of figures and conventions that enable works to have the forms and meanings they do. (1980, 8)

Although structuralism has been (and remains) a controversial movement, it has undoubted successes to its credit across a range of disciplines and subject matters. Within literary criticism its most unqualified successes have probably been within the field of NARRATOLOGY. Genette – one of its leading practitioners within this field – has argued that structuralism is more than just a method and needs to be seen as a general tendency of thought or an ideology (1982, 11). Interestingly, Genette sees the concern of structuralism with form at the expense of content as a *corrective* one, as can be seen from his (much-quoted) comment that 'Literature had long enough been regarded as a message without a code for it to become necessary to regard it for a time as a code without a message' (1982, 7).

Structure The concept of structure is not limited to STRUCTURALIST theory. The term is in common use in literary-critical discussion, and as one might expect, it is used in a number of relatively loose senses. It is usual to make a distinction between a literary work's structure and its plot (see STORY AND PLOT): whereas a work's plot can be seen as the NARRATIVE arrangement of its story, *structure* refers to its total (or total aesthetic) organization.

The term tends to be given a SYNCHRONIC, non-historical force, mainly as a result of the influence of Structuralist theories which see historical change in terms of the successive replacement of structures (rather than of the modification or development of a structure).

A more technical definition developed both from Structuralism and Systems Theory is given by Anthony Wilden: 'Structure is the ensemble of laws which govern the behavior of the system' (1972, 242). Moreover, these laws control elements or components which are interchangeable – thus an economic system stays the same even though the economic acts which it enables and controls are all unique. This takes us to the Structuralist commitment to a set of enabling rules ('LITERARINESS') which remain the same even though the literary works or acts of reading enabled or controlled by the rules change.

The paired terms *deep structure* and *surface structure*, originated by the Linguistician Noam Chomsky, denote elements in his Standard Theory. According to this theory, the following sentences would be seen to have the same deep structure, but different surface structures:

The ploughman homeward plods his weary way
Weary, the ploughman plods his way homeward
The ploughman plods his weary way homeward

Surface structure is derived from deep structure by means of transformations: hence transformational grammar. Chomsky's theory is controversial amongst Linguisticians, and amongst non-Linguisticians (Literary Critics, for example) it seems fair to say that the terms *deep structure* and *surface structure* are used only metaphorically.

See also PRAGUE SCHOOL.

Structure in dominance A term popularized in English by the writings of Louis Althusser, especially his essay 'On the Materialist Dialectic', which is to be found in *For Marx* (1969). Althusser uses the term to introduce a hierarchical element into the set of contradictions which, according to him, constitute a structured unity. This can be seen as a way of escaping one of the problematic elements inherent in the limitation to the SYNCHRONIC which characterizes STRUCTURALISM and neo-structuralism. Because the synchronic by definition excludes development, and because developmental change is the classic way of distinguishing determining from non-determining (or dominant from subservient) influences and forces, structuralism faces the problem that although it can trace relationships it lacks the ability to rank these relationships according to their importance (varyingly defined). The concept of structure in dominance can thus be seen as a reintroduction of the historical to an essentially synchronic schema by Althusser: dominance can only be asserted or revealed on a temporal continuum.

Structures of feeling The term is the coinage of the Welsh CULTURAL theorist and novelist Raymond Williams, who devotes a single chapter of his book *Marxism and Literature* to it. Williams introduces his discussion by arguing against a reduction of the social to 'fixed forms' and against its separation from the personal (1977, 128, 129). Thus Williams distances himself from what he sees as a tendency in more traditional Marxism to take 'terms of analysis as terms of substance' (1977, 129), and thus to deal with lived experience at at least one remove.

According to Williams, the term is deliberately chosen to emphasize a distinction from more formal concepts such as *world-view* or IDEOLOGY; the reason for this, he states, is that we are 'concerned with meanings and values as they are actively lived and felt', and with 'characteristic elements of impulse, restraint, and tone; specifically affective elements of consciousness and relationships: not feeling as against thought, but thought as felt and feeling as thought: practical consciousness of a present kind, in a living and interrelating continuity' (1977, 132).

The term, Williams adds, represents a cultural hypothesis (1977, 132), and it can be seen as one of a number of attempts to retain Marxism's analytical socio-historical approach while taking it closer to the ways in which people actually experience their lives.

Style and stylistics The noun *style* has a long history and wide set of meanings: the OED devotes over six page-columns to its various definitions. It derives from a Latin term meaning stake, or pointed instrument for writing (it shares a broad etymology with *stylus*), and modern meanings involve METAPHORICAL and METONYMIC extensions of this meaning. It is the thirteenth of the OED's definitions that is most relevant to our present concerns: 'The manner of expression characteristic of a particular writer (hence of an orator) or of a literary group or period; a writer's mode of expression considered in regard to clearness, effectiveness, beauty, and the like.'

In the course of the present century Stylistics has grown up as a recognized academic discipline, situated on the borderline between the study of language and the study of literature (although stylistic analysis can be and is applied to non-literary TEXTS), and concerned to engage in technical study and analysis of what the OED calls 'manners of expression'. It is important for students of literature because – as a result of its close relation to and impingement on Literary Studies – it has brought a number of more specifically linguistic terms and methods of analysis into literary criticism and theory.

Styles can be categorized according to a number of principles: deliverer's intention (a humorous style); receiver's evaluation (an imprecise style); context (an inappropriate style or REGISTER); aesthetic (an ornate style); level of formality (a colloquial style); social class (an urbane style) – and so on. For a Linguistician these are relatively imprecise categories, of course, and the academic study of style typically involves an attempt to analyse what are perceived impressionistically to be distinctive styles in more formal and objective ways, often through the use of statistical analysis directed at syntax, vocabulary, grammar, and so on. At this level Stylistics involves an attempt to back up the hunches of common READERS ('Hemingway has a distinctive, plain style') with statistical evidence. At another level, Stylistics can involve an attempt to go beyond the hunches of the common reader, detailing significant stylistic differences which may be functional but which are not necessarily noticed by the reader or listener.

At this point the question of purpose must be confronted. According to Geoffrey Leech and Michael Short's *Style in Fiction* (1981), we normally study style, 'because we want to explain something, and in general, literary stylistics has, implicitly or explicitly, the goal of explaining the relation between language and artistic function' (1981, 13). This of course is a central concern of the traditional discipline of *Rhetoric*, and in many ways Stylistics inherits and develops fundamental concerns of rhetoricians throughout the ages.

Stylisticians have often gone outside the Anglo-American literary-critical tradition for theoretical ideas and approaches. The RUSSIAN FORMALISTS, the PRAGUE SCHOOL, the circle around the Russian Mikhail Bakhtin, the Swiss Linguistician Ferdinand de Saussure, the German-American Leo Spitzer – the ideas of these have perhaps been of more interest to stylisticians than has the work of most Anglo-American literary critics: possibly because of the different trajectories taken by the academic study of literature in Britain and the United States on the one hand, and Continental Europe on the other. What is more, Anglo-American literary critics have frequently expressed reserve or scepticism about the possibility of the 'objective' analysis of literary texts producing results relevant to literary criticism: a good example is the chapter entitled 'Jakobson's Poetic Analyses' in Jonathan Culler's *Structuralist Poetics* (1975).

It is an interesting fact that the contribution of Stylistics to the literary-critical study of prose has been far less controversial, and more influential among literary critics, than has its contribution to the analysis of poetry. Even the most traditionalist of literary critics could hardly fail to find much of value and little to take objection to in Leech and Short's *Style in Fiction*; in contrast, Roman Jakobson's analyses of poetry have generally been taken by literary critics much as Culler takes them: as valuable demonstrations of what stylistic analysis cannot do. It seems that formal and statistical approaches are more suited to the analysis of NARRATIVE technique than they are to poetic expression – perhaps because many decisions that the writer of narrative has to make involve, at one level, choice from a relatively finite set of alternatives.

Style indirecte libre See FREE INDIRECT DISCOURSE

Styleme See SEMEME

Subaltern See MARGINALITY

Subject and subjectivity The traditional sense of *subject* as an abbreviation of *the conscious or thinking subject*, meaning the self or ego (OED), or individual *cogito*, has been pressed into service in a largely pejorative sense in recent theoretical writing. The main targets for attack in this process have been (i) the view that the human subject is somehow a point of origin for larger historical, social or even personal movements and events, and (ii) the belief that the individual human being is possessed of valid self-knowledge and is self-actuating – in a phrase, in charge and control of him/herself.

A more detailed account of one particular example of this use of *subject* can be found in the entry for INTERPELLATION, in which the position of the French MARXIST philosopher Louis Althusser is outlined. According to Althusser, all IDEOLOGY '*hails or interpellates concrete individuals as concrete subjects*, by the functioning of the category of the subject' (1971, 162). Thus *subject* represents the individual's self-consciousness and consciousness of self after having been 'body-snatched' by ideology. Althusser's position here forms the theoretical basis for a discussion by Etienne Balibar and Pierre Macherey of the specific rôle played by literature in this process. According to them, through the endless functioning of its TEXTS,

> literature unceasingly 'produces' *subjects*, on display for everyone. So paradoxically using the same schema we can say: literature endlessly transforms (concrete) individuals into subjects and endows them with a quasi-real hallucinatory individuality. (1973, 10)

These subjects are not just the Readers of literature, but also the Author and his Characters (Balibar and Macherey's capitals). The argument appears in part to be that as subjects are always opposed and seen in relation to objects and things, the impression is given that the motor or guiding element in history is the individual self or consciousness, rather than extra- or supra-individual forces. (The subject is active and living: external forces are dead and passive.)

It can be seen that this to a certain extent reflects a traditional Marxist subordination of the rôle of the individual to extra-individualistic forces. This same form of subordination can also be seen in obituaries for the death of the AUTHOR, except that in the case of Barthes's argument in 'The Death of the Author' it is language rather than the class struggle or the forces of production which is the extra-individual element involved. The individual reappears in this essay in the guise of the READER, however; the dissolution of the subject is perhaps more complete and more consistent in Michel Foucault's 'What is an Author?', and in this essay Foucault stresses that 'it is a matter of depriving the subject (or its substitute) of its role as originator, and of analyzing the subject as a variable and complex function of discourse' (1980b, 158).

Put another way: the subject is SITE rather than CENTRE or PRESENCE, is where things happen, or that to which things happen, rather than that which makes things happen: extra-individual forces use the subject to exert their sway, the subject does not use them (although it thinks that it does, and this is part of the cunning of the system). Compare, for example, Jonathan Culler's suggestion that as the self is broken down into component systems and is deprived of its status as source and master of meaning, it comes to seem more and more like a construct: 'a result of systems of convention', such that 'even the idea of personal identity emerges through the discourse of a culture: the "I" is not something given but comes to exist as that which is addressed by and relates to others' (1981, 33).

This is the stance generally adopted by POST-STRUCTURALISM, for which a major target is the view of the subject as primary, unified, self-present, self-determining, autonomous, and homogeneous. For post-structuralism the subject is secondary, constructed (by language, or ideology, for instance), volatile, standing in its own shadow, and self-divided. An important influential theorist here is the French psychoanalyst Jacques Lacan, whose writing on the MIRROR-STAGE is much quoted by subsequent theorists concerned to explore the creation of the subject from a rather different angle from that adopted by Althusser.

If we turn to recent FEMINIST theory we find a rather more nuanced attitude to the subject and to subjectivity. During the earlier years of the rebirth of the Women's Movement in the 1960s and 1970s one finds evidence of a far less antagonistic view of the subject – a belief that the subjective might actually provide a rallying point *against* SEXIST ideas, and against the ideology of PATRIARCHY. In June 1971, for example, Doris Lessing wrote a new *Preface* to her novel *The Golden Notebook* – a novel which from its first publication in 1962 had been extremely influential in what became known as the Women's Liberation Movement. It is noteworthy that Lessing devotes a lot of attention to the issue of subjectivity in her comments on the novel.

> When I began writing there was pressure on writers not to be 'subjective'. This pressure began inside communist movements, as a development of the social literary criticism developed in Russia in the nineteenth century, by a group of remarkable talents, of whom Belinsky was the best known . . . (1973, 12)

Lessing notes, however, that alongside this pressure on the writer to ignore 'stupid personal concerns when Rome is burning', novels, stories, art of every sort, became more and more personal (1973, 13). And finally, she reports, she became aware that the way out of the dilemma with which this confronted the writer was to recognize that 'nothing is personal, in the sense that it is uniquely one's own', and that the way to deal with the problem of subjectivity was to make the personal general by seeing the individual as a microcosm (1973, 13).

Lessing's position here is representative of a widespread tendency in the women's movement in the 1960s and 1970s, a tendency to consider (especially a woman's) subjective feelings, responses, beliefs as reliable and worthy of being encouraged and nurtured as a force oppositional to patriarchy.

See also INTERSUBJECTIVITY.

Sub-text That which is implied but not directly or overtly stated. The term originates in theatrical usage, and is associated with the so-called *Theatre of Silence* – the normal English translation of the *Théâtre de l'Inexprimé* founded by Jean-Jacques Bernard in the 1920s. The modern dramatist most associated with the term is Harold Pinter, whose plays typically have, critics agree, sub-texts of a violent or sexual nature which is unstated on the surface. The term typically also implies a certain consistency in implied MEANING: thus the sub-text of a given work is unlikely to consist of a sequence of meanings which have nothing in common with one another.

The term has achieved wide usage in recent years, partly because of its points of contact with the influential theory of conversational implicature contained in SPEECH ACT THEORY, and partly because it chimes in with a range of critical theories which, in various ways, argue that literature in particular and language in general often function by means of indirect, hidden, or implied meanings as much as by overt and direct ones.

A related but somewhat different theoretical term that has achieved a much more restricted circulation is that of the *suggestiveness* of literary works. Krishna Rayan's book *Text and Sub-text: Suggestion in Literature* (1987) attempts to establish a link between the two terms.

It should be pointed out that a sub-text is the creation (conscious or unconscious) of the AUTHOR; it can be discovered by READER or audience, but not created by them. But at the heart of the theory of suggestiveness is the idea that the writer encourages the reader to bring his or her creativity into play, and by textual means indicates a direction in which the reader or audience may proceed to explore unstated possibilities. Suggestion, in brief, offers a possible way whereby reader- or audience-creativity may be theoretically justified and LEGITIMIZED. This is not to say that the author cannot influence the way in which a reader responds to suggestive elements in a work (as the more specifically sexual meaning of the term *suggestive* makes clear, and as Laurence Sterne demonstrated in his *Tristram Shandy*).

Suggestiveness See SUB-TEXT

Summary See DURATION

Super-ego See TOPOGRAPHICAL MODEL OF THE MIND

Superstructure See BASE AND SUPERSTRUCTURE

Supplement See MARGINALITY

Surface structure See STRUCTURE

Surfiction See MODERNISM AND POSTMODERNISM

Suspense The everyday meaning of this word has long been applied to the reading of literature, especially fiction, and to the experience of the spectator of drama. Being in a state of heightened excitement as a result of wanting to know what happens next and what will happen ultimately is one of the forces that keeps the READER turning the pages (or glues the audience to its seats).

Clemens Lugowski has suggested a useful distinction between 'result-oriented suspense' and 'process-oriented suspense' (1990, 37). The former is where the reader/audience is anxious about *what* will happen; the latter where the anxiety is directed towards *how* it will happen. There was an amusing discussion of the distinction between different sorts of suspense in a Tony Hancock TV show from 1960 called 'The Last Page'. Hancock is reading a crime novel entitled *Lady Don't Fall Backwards* by D'Arcy Sarto, and explains to Sidney James how all Sarto's stories end.

> **Tony** He always keeps you in suspense till the last page, this bloke. (*Reads on, getting excited*) Yes, I thought so, he's invited everybody into his flat. He always does that. He lashes them up with drinks, lights a cigarette and explains who did it. Then the murderer rushes to the window, slips and falls, hits the pavement, and Johnny Oxford turns round to the guests, finishes his Manhattan and says, 'New York is now a cleaner place to live in.' The End. Turn over, a list of new books and an advert for skinny blokes.
> **Sidney** If you know what's going to happen every time, why bother to read it?
> **Tony** Because I don't know who it is who's going to hit the pavement.
> (Galton & Simpson 1987, 98)

As can be seen, the creation of suspense is often connected to manipulation of FORMULAIC possibilities.

In the realm of POPULAR art and literature, the word 'suspense' has assumed a generic force, and shops hiring out video films often use the word as a particular category.

Suture The technical term for the stitching up or joining-together of the lips of a wound has been applied by Steven Cohan and Linda M. Shires to the production of the SUBJECT. According to them the subject is as it were 'stitched' to DISCOURSE by the SIGNIFIER (1988, 162). They base their argument on the expansion of a comment of Jacques Lacan's by Jacques-Alain Miller. Lacan had referred to suture as the conjunction of the imaginary and the symbolic; Miller develops this suggestion by arguing that suture names the relation of the subject to the chain of its discourse, in which it is said to figure as an element lacking, or stand-in. Cohan and Shires in turn consider the relevance of these ideas to NARRATIVE, and they suggest that reading or viewing narrative 'involves the continuous suturing of a narrated subject whose pleasure is secured, jeopardized, and rescued by a signifier' (1988, 162).

Sometimes applied to the functioning of IDEOLOGY, which is continually splitting opening and (the argument goes) having to be sutured back together by HEGEMONIC practices.

Switchback See ANALEPSIS

Syllepsis A cluster or gathering (of events, circumstances, experiences, etc.) ordered according to some principle other than temporal unity or sequence. 'John told her all the things that had happened in his mother's house, and she recounted all the times that she had visited a boy-friend's parents' contains two syllepses: one based on situational coherence, the other on thematic coherence.

Symbolic See IMAGINARY / SYMBOLIC / REAL

Symbolic code See CODE

Symptomatic reading See READERS AND READING

Synchronic See DIACHRONIC AND SYNCHRONIC

Synecdoche See SYNTAGMATIC AND PARADIGMATIC

Synonymous characters A concept suggested by Mieke Bal, and related to her proposal that one distinguish between a CHARACTER's relevant characteristics and his or her secondary characteristics through a mapping of these characteristics on suitable SEMANTIC AXES. Those characters who end up with exactly the same positive, negative, and unmarked elements after such a mapping are deemed to be synonymous. The concept has much in common with the more traditional term *type*, except that synonymous characters share characteristics which may be specific to a particular WORK whereas types normally have a CONVENTIONAL significance which goes beyond the confines of an individual work. Thus to suggest that characters in different works are synonymous one needs to use a term such as type or FUNCTION (function in the sense in which a writer such as Vladimir Propp uses it, although Propp focuses upon 'action-by-a-character' rather than upon character as such).

Syntagm See SYNTAGMATIC AND PARADIGMATIC

Syntagmatic and paradigmatic According to Saussure's account of language in the *Course in General Linguistics*, '[c]ombinations supported by linearity are *syntagms*', while those 'co-ordinations formed outside discourse' that are 'not supported by linearity' but are 'a part of the inner storehouse that makes up the language of each speaker', 'are *associative relations*' (1974, 123). Present-day usage tends to favour the term *paradigms* over *associative relations*. The distinction can perhaps best be explained by means of a particular example. To construct a grammatical sentence we have to select words according to one set of rules, and combine them according to another set. The first are *paradigmatic* (or *associative*) rules and the second are *syntagmatic* rules. Thus in the sentence

> The cat sat on the mat

The first word could be replaced by 'A' or 'No'; the second word could be replaced by 'dog' or 'boy'. The relations between these groups of interchangeable words (interchangeable according to the rules of grammar and syntax, but not, of course, *semantically* interchangeable) are *paradigmatic*; that is to say, they involve rules which govern the *selection* (not the combination) of words used in a sentence. But once one has chosen 'The' as the first word, one's selection of the second word is constrained: it cannot, for example, be 'a'. This is because there are rules governing the *combination* of words in a sentence: *syntagmatic* rules.

Another way of expressing the distinction is as Jonathan Culler puts it: syntagmatic relations bear on the possibility of combination; paradigmatic relations determine the possibility of substitution (1975, 13).

In 1956 Roman Jakobson published an article entitled 'Two Aspects of Language and Two Types of Aphasic Disturbances' which relied heavily upon a distinction between what Jakobson referred to as the metaphoric and the metonymic modes. In traditional usage *metonymy* is a figure of speech in which the name of one item is given to another item associated by *contiguity* to it. Thus 'The pen is mightier than the sword' works by means of metonymy: *pen* and *sword* stand for those activities with which they are closely associated. Under the heading of 'metonymy' Jakobson includes *synecdoche* – that is, the use of a part to represent a whole: 'bloodshed' for war, for example. *Metaphor*, in contrast, relies upon *similarity* rather than contiguity. Thus 'Eddie the eagle' (a strikingly unsuccessful British ski-jumper) was so-called not because he was often in the company of eagles, but because it was suggested (with a generous helping of irony) that a similarity could be observed between his ski-jumping and the flight of an eagle.

Through reference to the varying forms of aphasia experienced by patients with cerebral impairments, Jakobson was able to show that the metaphoric and metonymic processes were governed by localized brain functions. These different processes (or modes) Jakobson related, in turn, to what he described, following Saussure, as the selection and combination axes of language. Metaphor and metonymy were thus, according to Jakobson, governed by specific brain functions which also governed these two fundamental axes of language. According to Jakobson, all examples of aphasic disturbance consisted of some impairment of either the faculty for selection or of that for combination and contexture.

> The former affliction involves a deterioration of metalinguistic operations, while the latter damages the capacity for maintaining the hierarchy of linguistic units. The relation of similarity is suppressed in the former, the relation of contiguity in the latter type of aphasia. Metaphor is alien to the similarity disorder, and metonymy to the contiguity disorder. (Jakobson & Halle 1971, 90)

Jakobson did not stop here, but attempted to generalize his discoveries and to claim that although both processes were operative in normal verbal behaviour, the influence of cultural pattern, personality or verbal style could lead to preference being given to one of the two processes over the other. From here, Jakobson then moves to consider literature. He claims that whereas the metaphoric process is primary in Romanticism and Symbolism, metonymy is predominant in REALISM (Jakobson & Halle 1971, 91–2).

Jakobson's article has itself been very influential; Jacques Lacan refers to it in his essay 'The Agency of the Letter in the Unconscious or Reason Since

Freud' (in Lacan, 1977, 146–78), and David Lodge structures his book *The Modes of Modern Writing* (1977) around an attempt to apply and extend what Jakobson says about metaphor and metonymy to the analysis and interpretation of literary TEXTS.

In an essay entitled 'The Semantics of Metaphor' Umberto Eco has suggested that metaphor and metonymy are linked at a deeper level, claiming that 'each metaphor can be traced back to a subjacent chain of metonymic connections which constitute the framework of the code and upon which is based the constitution of any semantic field, whether partial or (in theory) global' (1981, 68).

According to Robert Scholes, 'neo-Freudians' such as 'Jacques Lacan and his circle' have reminded us that Jakobson's metaphor and metonymy are very close in meaning to Freud's CONDENSATION AND DISPLACEMENT (1982, 75–6).

System Unlike the term STRUCTURE, *system* tends to be given a number of varied meanings in current theoretical discussion. Perhaps its most precise meaning is to be found in structural Linguistics: here language is seen as a system of relationships based on DIFFERENCE and capable of generating MEANINGS.

Apart from this relatively precise usage the term can be used more or less synonymously with the STRUCTURALIST term *structure* – somewhat confusingly, as a system is typically characterized by its dynamic, goal-seeking impetus, whereas a structure is more usually seen as a succession of over-arching, constant rules (see the separate entry for this term). As Josué Harari puts it, in a discussion of Lévi-Strauss's structural anthropology, 'history' tends to be replaced by 'diachrony', with the former falling under the tutelage of the '*System*' – rather than, we presume, *structure* (1980, 20).

Doubtless it is for this reason that Robert Young insists that for Roland Barthes, *codes of reading* (see CODE) do not constitute 'a rigorous, unified system': they operate simply as 'associative fields', a supra-textual organisation of notations which impose a certain idea of structure' (1981, 134).

In Michel Foucault's writing a discipline (in the sense of an academic discipline) is seen as 'a sort of anonymous system at the disposal of anyone who wants to or is able to use it' (1981, 59).

For Michael Riffaterre, a literary work's system is 'a network of words related to one another around a central concept embodied in a kernel word' (1981, 114).

T

Technological determinism The belief that technological changes and innovations necessarily (and often mechanically) lead to changes in society, CULTURE, and on occasions art. Such a position is normally associated with the belief that all such changes start with technological developments, or (to put it the other way round) that all changes in society, culture or art can be traced back to technology. In the work of Marshal McLuhan, who has been accused of espousing technological determinism in his work, one can also find a related naïvety concerning, for example, the force and influence of economic and political factors.

In the field of Media Studies technical determinism is often relatively easy to expose: many leaps forward on the technological plane (such as 'talkies' or TV) have to wait until the right social, political and (not least) economic factors allow their implementation.

Telos See CENTRE

Tempo See DURATION

Tense See LINGUISTIC PARADIGM

Terrorism A term coined by the French writer Jean Paulhan, and made use of by Gérard Genette, to describe a type of writing which refuses to make use of the 'flowers of rhetoric' and which entails a refusal of any of the traditional supports and devices of literature or *belles lettres* (Genette 1982, 60, n5).

Also used more pejoratively to suggest a method of argumentation that is unprincipled and aggressive (and often SEXIST). Thus Jean-François Lyotard, talking of 'terrorist behaviour', defines terror as

> the efficiency gained by eliminating, or threatening to eliminate, a player from the language game one shares with him. He is silenced, or consents, not because he has been refuted, but because his ability to participate has been threatened (there are many ways to prevent someone from playing). (1984, 63–4)

Tessera See REVISIONISM

Text and work In his essay 'Theory of the Text' Roland Barthes suggests that whereas the work is 'a finished object, something computable, which can occupy a physical space', 'the text is a methodological field', and that 'The work is held in the hand, the text in language'. Barthes continues, suggesting that one can put the matter in another way:

> [I]f the work can be defined in terms that are heterogeneous to language (everything from the format of the book to the socio-historical determinations which produced that book), the text, for its part, remains homogeneous to language through and through: it is nothing other than language and can exist only through a language other than itself. In other words, 'the text can be felt only in

> a work, a production': that of 'signifiance'. (1981b, 39–40; Barthes's quotation is from his own *Image-Music-Text*. For *signifiance*, see the entry for SIGN)

This suggests a different definition of *work*, at any rate, from that current in most contemporary literary-critical usage, in which the term applies to the literary composition seen independently of its particular physical manifestations (*Hamlet* the work would still exist even if there were no physical copies of it so long as there were anyone who could recall the words that comprise it and pass these on to others).

In practice, these two terms have come to enjoy a rather different usage in recent years: *text* has become the preferred term for referring to a literary or other work (not necessarily linguistic or verbal) stripped of traditional preconceptions about autonomy, authorial control, artistic or aesthetic force, and so on. Thus use of the term text and its cognates in contemporary literary-critical circles is often associated with an attempt to argue that the traditional distinction between literary and non-literary texts should be either made less absolute, or scrapped altogether.

Text Linguistics forms a sub-section of the discipline of Linguistics, and is very close to discourse analysis (which, as the entry for DISCOURSE argues, is itself a term with no clear single meaning). Michael Stubbs (1983) treats text and discourse as more or less synonymous, but notes that in other usages a text may be written while a discourse is spoken, a text may be non-interactive whereas a discourse is interactive (see the quotation from Leech and Short below), a text may be short or long whereas a discourse implies a certain length, and a text must be possessed of surface cohesion whereas a discourse must be possessed of a deeper coherence. Finally, Stubbs notes that other theorists distinguish between abstract theoretical construct and PRAGMATIC realization, although, confusingly, such theorists are not agreed upon which of these is represented by the term *text*.

Text Linguistics is generally taken to be broader than discourse analysis, however, and is normally taken to include, among other things, elements from STYLISTICS and NARRATOLOGY. Geoffrey Leech and Michael Short argue for one way of distinguishing between text and discourse that has already been mentioned:

> Discourse is linguistic communication seen as a transaction between speaker and hearer, as an interpersonal activity whose form is determined by its social purpose. Text is linguistic communication (either spoken or written) seen simply as a message coded in its auditory or visual medium. (1981, 209)

The key phrase here is clearly 'seen simply as a message', an emphasis which is contrasted with 'a transaction between speaker and hearer'. This suggests that the identical words can both constitute a text, if seen simply as a message, but can also be seen as an element in a discourse if seen in terms of their mediating a transaction between speaker and hearer. As we have already noted, this implies that to talk of a text is to concentrate on the language used and to ignore or play down the context in which it is used. The problem comes when one tries to define what is meant by 'seen simply as a message', for some theorists would argue, not illogically, that to see something as a message (even 'simply') one has to posit some sort of a context, for a message is defined as a message not just by its linguistic characteristics, but by the message *function*. 'Get lost' is a message only when used to convey information from a source to a destination: when used to illustrate English spelling it is not a message. (The OED starts its definition of *message* as follows: 'A communication transmitted through a messenger or other agency; an oral or written communication sent from one person to another.' This sounds very like 'a transaction between speaker and hearer'.)

Katie Wales quotes the seven criteria for textuality given by de Beaugrande and Dressler in their *Introduction to Text Linguistics* (1981): cohesion and coherence, intentionality, acceptability, situationality, informativity, and intertextuality (1989, 459). These certainly offer a relatively workable definition, and might enable one to use the term *work* as a more specialized sub-set of *text*.

Textualist According to Richard Rorty,

> In the last century there were philosophers who argued that nothing exists but ideas. In our century there are people who write as if there were nothing but texts. (1982, 139)

These last Rorty dubs 'textualists'. In his list of textualists Rorty includes the 'so-called "Yale School" of literary criticism', which he claims centres around Harold Bloom, Geoffrey Hartmann, J. Hillis Miller, and Paul de Man; POST-STRUCTURALIST French thinkers such as Jacques Derrida and Michel Foucault, historians such as Hayden White, and social scientists such as Paul Rabinow.

What all of these individuals have in common, according to Rorty, is (i) an antagonistic position to natural science and (ii) the belief that we can never compare human thought or language with 'bare, unmediated reality' (1982, 139). Rorty sees these positions as constituting a textualism which is the contemporary counterpart of idealism, and its practitioners as the spiritual descendants of the idealists (1982, 140).

Theme and thematics In traditional literary-critical usage the term *theme* suffers from a certain ambiguity. Whereas for some critics it implies a certain claim, doctrine or argument raised either overtly or implicitly throughout a literary WORK (or by the work as a whole), for others the term *thesis* is used with this meaning, and theme is reserved for what one can better call an issue. Prince suggests the useful distinction that whereas a thesis involves both a question and a proposed answer or answers, a theme 'does not promote an answer but helps to raise questions' (1988, 97). Disagreement can also be found as to whether thematic elements in a literary work are dependent upon an AUTHOR's conscious and deliberate attempt to raise certain issues, or whether thematic elements can be raised, as it were, unconsciously. A theme is generally also distinguished from a *motif* (or *leitmotif*) by its greater abstraction.

Prince also relates theme to FRAME, suggesting that a theme has a framing function in a literary work (1988, 111).

To show how shifting is the use of these different terms, one can note that M. H. Abrams suggests that motif and theme are sometimes used interchangeably, but that theme is more usefully applied to a general claim or doctrine (1988, 111) – which of course is what Prince would dub a thesis.

Some theorists have adopted the term *thematics* in recent years, especially in the study of NARRATIVE. Thematics is distinguished from narrative on a theoretical level although the two are interconnected in practice, narrative technique being one of the ways in which thematics are generated and modified. Thus thematics tends to refer to an end-product: it is the sum of issues raised, normally expressed in a hierarchy of questions and problems, with perhaps some suggested answers. It is not necessarily dependent upon the author's conscious or unconscious intentions.

According to Mieke Bal, *thematic space* refers to a described space in a literary work which assumes a thematic function. In other words, it is not just a SITE on which actions take place, but 'an "acting place" rather than the place of action' (1985, 95).

Thesis See THEME AND THEMATICS

Thick description / thin description The person usually credited with coining this term is the anthropologist Clifford Geertz, but in his 'Thick description: Toward an Interpretive Theory of Culture', the first chapter of *The Interpretation of Cultures* (1973), he attributes the term to two essays by the philosopher Gilbert Ryle: 'Thinking and Reflecting' and 'The Thinking of Thoughts'. Both are reprinted in the second volume of Ryle's *Collected Papers* (1971).

Ryle contrasts the behaviour of two boys, both of whom are rapidly contracting the eyelids of the right eye.

> In the first boy this is only an involuntary twitch; but the other is winking conspiratorially to an accomplice. At the lowest or thinnest level of description the two contractions of the eyelids may be exactly alike.
>
> . . .
>
> [The winker acknowledges] that he had not had an involuntary twitch but (1) had deliberately winked, (2) to someone in particular, (3) in order to impart a particular message, (4) according to an understood code, (5) without the cognisance of the rest of the company . . . (Ryle 1971, 481)

But Ryle adds another stage of possibility: that of a third boy who to give amusement to his cronies, deliberately parodies the first boy's behaviour with a wink of his own. We now have three different 'winks' which, in purely physical terms, are identical, but which in a wider sense are clearly very different. According to Ryle a *thin description* of the winks would see them as identical, whereas a *thick description* of them would recognize their variety. Geertz argues that between such a thin description and such a thick description

> lies the object of ethnography: a stratified hierarchy of meaningful structures in terms of which twitches, winks, fake-winks, parodies, rehearsals of parodies are produced, perceived, and interpreted, and without which they would not (not even the zero-form twitches, which, *as a cultural category*, are as much nonwinks as winks are nontwitches) in fact exist, no matter what anyone did or didn't do with his eyelids. (1973, 7)

According to Geertz, 'ethnography is thick description' (1973, 9–10).

It is perhaps the development of NEW HISTORICISM that has done most to popularize the use of the term in conjunction with literary studies. Brook Thomas has suggested that in certain New Historicist hands, thick description has degenerated into a sort of ORGANICISM that believes in the possibility of reading of (presumably social or historical) whole from a (very isolated) part (1991, 10), and he suggests that to demand a thick description of the symbolic world of goods is 'to open up vistas of interpretation that are almost vertiginous in their potential complexity' (1991, 11–12).

Top down (perspective) See SOLUTION FROM ABOVE / BELOW

Topic Usefully defined by Umberto Eco as the textual operator which is needed to realize all of the relevant semantic disclosures in a DISCURSIVE structure (1981, 23).

Eco contrasts this term with Greimas's term *isotopy*, quoting Greimas's definition of this as 'a redundant set of semantic categories which make possible the uniform reading of the story' (1981, 26). The distinction, for Eco, is that whereas the topic governs the semantic properties that can or must be taken into account during the reading of a given TEXT, an isotopy is the actual textual verification of that hypothesis which the topic produces. Put another way: a topic leads the READER to have certain expectations, whereas an isotopy is a READING based upon these expectations.

Topographical model of the mind Topographical means representation by means of a map: topographical models of the mind represent the mind's structure through a spatial model which 'maps' its different elements by assigning them different mental 'spaces'. (Note: not different *physical* spaces: one must remember that 'mind' is not the same as 'brain'.)

The best-known of such models is Sigmund Freud's division of the mind into *id* (those instinctual drives arising from the BODY's constitutional needs; *ego* (that agency deriving from and regulating the id and the instinctual drives);

and *super-ego* (the mental transformation of social/parental influences on and modifications of the instinctual drives). This particular topography comes later in Freud's career, and replaces a number of earlier models. Jacques Lacan has referred to his own DISCOURSE in which 'each term is sustained only in its topological relation with the others' (1979, 89) – a classic example of the use of the topographical model and one which perhaps betrays the potential dangers of self-enclosure that its use often seems to involve.

This and other models have been used by PSYCHOLOGICAL and psychoanalytic literary critics in various ways to 'map' aspects of – for example – the READER's experience of a TEXT.

Totalizing discourse Any DISCOURSE which seeks to occupy all the available ground and thus deny any oppositional SITE to those whom it excludes.

Trace See ARCHE-WRITING

Traditional intellectuals See INTELLECTUALS

Transcendental pretence / signified / subject The influence of Jacques Derrida has radically changed the connotations of the word *transcendental* and its cognates for many. Whereas at one time its associations were mainly positive – 'that which is above all other categories of thing' – since Derrida the word is associated with a belief in fixed, extra-linguistic points of meaning-determination, a view which he characterizes as LOGOCENTRIC and representative of the METAPHYSICS OF PRESENCE. He points out, incidentally, that those who consider themselves to be MATERIALISTS are just as liable to treat 'matter' as a transcendental signified as those believing themselves idealists are liable to find their own transcendental signified in God or whatever (1981b, 65).

According to Derrida himself, from his very first published work he sought to 'systematize a deconstructive critique precisely against the authority of meaning, as the *transcendental signified* or as *telos*, in other words history determined in the last analysis as the history of meaning, history in its logocentric, metaphysical, idealist . . . representation' (1981b, 49–50). According to Alex Callinicos, Derrida believes that

> any attempt to halt the endless play of signifiers, above all by appealing to the concept of reference, must . . . involve postulating a 'transcendental signified' which is somehow present to the consciousness without any discursive mediation. (1989, 74)

The implications of such a position for literary criticism are not far to seek: the TEXT is also subject to a totalizing play of linguistic DIFFERENCE which cannot be fixed or organized by any extra-systemic reference-point – AUTHOR, authorial intention, 'common reader's' interpretation, or whatever.

Analogous objections are raised against the transcendental SUBJECT – that is, against the belief that the *ego* is undetermined by and independent of social and CULTURAL forces, and that it constitutes a unity rather than a SITE for the play of contradictions. The transcendental pretence, according to Robert C. Solomon, rests on an IDEOLOGICAL belief that the 'white middle classes of European descent were the representatives of all humanity, and that as human nature is one, then so too must its history be one as well' (1980, xii).

Finally, in *Positions* Derrida points out that it is also possible to believe in a transcendental signif*ier*, and he gives as example the *phallus*, when seen 'as the correlate of a primary signified, castration and the mother's desire' (1981b, 86). (See the entry for PHALLOCENTRISM.)

Transcoding See MEDIATION

Transference In the work of Sigmund Freud, the transferring of feelings originally associated with an infantile object, childhood trauma or other object of psychoanalytic investigation, from its source to the investigating psychoanalyst. *Studies on Hysteria* provides an early example of Freud's development of the concept. A female patient of Freud's was possessed of a particular hysterical symptom which originated in a wish, suppressed because of the fear it caused, that a man with whom she had been talking would take the initiative and kiss her. At the end of a psychoanalytic session the wish re-emerged, with Freud as its object, as a result of which the patient was horrified, spent a sleepless night, and was 'quite useless for work' at the next session. But once the wish was uncovered psychoanalysis proceeded further, and the original, frightening, wish was then revealed (Freud & Breuer 1974, 390).

Counter-transference, in contrast, is where the analyst's own desires are imposed upon the person undergoing analysis.

In recent literary theory the concept of transference is sometimes broadened to include any process whereby the analyst of a TEXT becomes inextricably involved in the object of his or her analysis, such that distinguishing between what is 'in' the text and what has been put there by the analyst in the process of analysis, is impossible to determine. The interest of the concept to theorists of literature is in part that it provides a name for a well-known process: that whereby the analyst or interpreter of a literary WORK can change that work for him or herself and for subsequent readers.

Transgredient According to Tzvetan Todorov's *Mikhail Bakhtin: The Dialogical Principle* (1984), a term which Bakhtin borrowed from Jonas Cohn, *Allgemeine Ästhetik* (Leipzig, 1901). Todorov sees this as a term complementary to *ingredient*, 'to designate elements of consciousness that are external to but nonetheless absolutely necessary for its completion, for its achievement of totalization' (1984, 95).

See also EXOTOPY.

Transgressive strategy A term sometimes used by POST-STRUCTURALISTS to indicate any way of dealing with a TEXT which attempts to go beyond the assumptions upon which it is based and which (if not challenged) it will reproduce. Transgressive strategies are *denaturalizing* – that is, they prevent us from seeing the text *as* natural or *in* a 'natural' way (a way that does not go against the grain of CONVENTION).

See also OPPOSITIONAL READING.

Transparency reading / transparent criticism See OPAQUE AND TRANSPARENT CRITICISM

Transtextuality See INTERTEXTUALITY

Transworld identity See HOMONYMY

Uncanny See FANTASTIC

Uncle Charles Principle Monika Fludernik provides an economical account of this term:

> This phenomenon was not discovered in English criticism until Hugh Kenner's response to Wyndham Lewis's remarks on [James] Joyce's sentence: *Uncle Charles repaired to the outhouse* [from *A Portrait of the Artist as a Young Man*]. Lewis had criticized Joyce for slipping into authorial style (ruining the effect of immediacy of presentation, i.e. the reader's illusion of being within Stephen's consciousness throughout the novel). Kenner, by contrast, argues that *repair* in fact echoes Uncle Charles's linguistic preciousness, describing the action in the very terms he might euphemistically have used himself. Kenner then baptized such infection of the narrative by figural language the 'Uncle Charles Principle'. (Fludernik 1993, 332; she refers the reader to Kenner 1978: ch. 2)

See also FREE INDIRECT DISCOURSE (coloured narrative).

Unconscious Jacques Lacan defines the unconscious (he does not grant the term a capital letter) as 'that part of the concrete discourse, in so far as it is trans-individual, that is not at the disposal of the subject in re-establishing the continuity of his conscious discourse' (1977, 49), and as 'that chapter of my history that is marked by a blank or occupied by a falsehood: it is the censored chapter' (1977, 50), and it shows us 'the gap through which neurosis recreates a harmony with a real – a real that may well not be determined' (1979, 22).

Even so, it can be rediscovered – in monuments, in archival documents such as childhood memories, in semantic evolution, in traditions, and in surviving (conscious) traces (1977, 50). Moreover, the unconscious for Lacan is 'the sum of the effects of speech on a subject, at the level at which the subject constitutes himself out of the effects of the signifier' (1979, 126), and '*the discourse of the Other*' (1979, 131).

Lacan provides a rather more accessible account of the unconscious in his lecture 'Of Structure as an Inmixing of an Otherness Prerequisite to Any Subject Whatsoever', in which he insists that the unconscious has nothing to do with instinct or primitive knowledge or preparation of thoughts in some underground.

> I could see Baltimore through the window and it was a very interesting moment because it was not quite daylight and a neon sign indicated to me every minute the change of time, and naturally there was heavy traffic, and I remarked to myself that exactly all that I could see, except for some trees in the distance, was the result of thoughts, actively thinking thoughts, where the function played by the subjects was not completely obvious. In any case the so-called *Dasein*, as a definition of the subject, was there in this rather intermittent or fading spectator. The best image to sum up the unconscious is Baltimore in the early morning.

Anthony Wilden has drawn attention to Lacan's debt to the work of Lévi-Strauss in arriving at a linguistic model of the Unconscious (in which the Unconscious is seen to be structured like a language). Wilden further draws attention to the fact that Lévi-Strauss has said that his own development of a theory of the Unconscious was influenced both by Freud and Marx – and also by geology! Wilden argues that Lévi-Strauss reformulates the concept of the Unconscious as a locus 'not of instincts, not of phantasies, not of energy or entities, but as a locus of a *symbolic function* – a set of rules governing the possible messages in the system, a sort of syntax or code'. It was this formulation, Wilden argues, that allowed Lacan to declare first (in 1953) that 'The Freudian Unconscious is the discourse of the other', and shortly afterwards that 'The unconscious is structured like a language' (1972, 15).

Fredric Jameson's *The Political Unconscious* attempts to rehistoricize Freud's concept of the Unconscious and to reassign the CENTRE of the Freudian interpretive system (which Jameson sees as wish-fulfilment) to history and society rather than to the individual SUBJECT and individual psychobiology.

In their *Anti-Oedipus: Capitalism and Schizophrenia*, Gilles Deleuze and Félix Guattari remark that 'Women's Liberation movements contain, in a more or less ambiguous state, what belongs to all requirements of liberation: the force of the unconscious itself, the investment by desire of the social field, the disinvestment of repressive structures' (1983, 61). (For a critical FEMINIST view of Deleuze and Guattari, see the entry for DESIRE.)

See also ARCHE-WRITING; CONDENSATION AND DISPLACEMENT; PRIMARY PROCESS; TOPOGRAPHICAL MODEL OF THE MIND.

Unfolding A term used in the English translation of Roland Barthes's *S/Z*. As example Barthes suggests that 'to enter' can be unfolded into 'to appear' and 'to penetrate' (1990, 82). In other words, unfolding involves the semantic, CONNOTATIVE or IDEOLOGICAL unpacking of a term or word.

Unmotivated See ARBITRARY

Use value See FETISHISM

Uses and gratifications A term taken from Media Studies, associated especially with a family of theories concerning the way in which television audiences 'use' the programmes they watch to 'gratify' different needs or desires. The theory was developed in the United States and offered a view of TV audiences as more active than the pictures presented by both MARXIST critics and also by more populist 'prophets of doom' pronouncing on the evils of television.

Utterance An utterance is generally regarded as a natural unit of linguistic *communication.* In her editorial Preface to her translation of M. M. Bakhtin's *Problems of Dostoevsky's Poetics*, Caryl Emerson claims that the distinction between utterance and sentence

> is [Bakhtin's] own: a sentence is a unit of language, while an utterance is a unit of communication. Sentences are relatively complete thoughts existing within a single speaker's speech, and the pauses between them are 'grammatical,' matters of punctuation. Utterances, on the other hand, are impulses, and cannot be so normatively transcribed; their boundaries are marked only by a change of speech subject. (Bakhtin 1984, xxxiv)

In like manner, Jan Mukařovský attributes 'uniqueness and nonrepeatability' to the utterance in his essay 'The Esthetics of Language' (1964, 63).

It should be noted that an utterance can be either spoken or written (or, presumably, expressed in silent thought to oneself).

See also DISCOURSE; ÉCRITURE; ENUNCIATION; LANGUE AND PAROLE; and SPEECH.

Virtuality See PHENOMENOLOGY

Vision See PERSPECTIVE AND VOICE

Voice See PERSPECTIVE AND VOICE; POLYPHONY

Vorurteil A term from the German HERMENEUTIC tradition referring to the reader's predisposition to interpret in one way rather than another. Howard Felperin points out that whereas Martin Heidegger uses the term in the sense of 'pre-judgement', for Hans-Georg Gadamer it has more the force of 'prejudice' (1990, 110, n11).

Vraisemblance A French loan-word meaning 'appearing real', close to the English *verisimilitude*, and used in discussions of REALISM. The term is not new (its use in discussion of art dates from the seventeenth century), but what is new is that it is used in an increasingly pejorative or dismissive sense the more the status of realism has been brought into question.

Weak textualist See TEXTUALIST

Work See TEXT AND WORK

Writerly See READERLY AND WRITERLY TEXTS

Writing See ÉCRITURE

Z

Zero (degree writing) See ÉCRITURE

Zero focalization See PERSPECTIVE AND VOICE

Bibliography

This bibliography lists all works from which extracts are quoted, along with some other important works mentioned in entries.

Abrams, M. H. (1977). The limits of pluralism: the deconstructive angel. *Critical Inquiry*, 3, 425–38.

Abrams, M. H. (1988). *A Glossary of Literary Terms*. (5th edn.) London: Holt, Rinehart & Winston.

Ades, Dawn (1976). *Photomontage*. London: Thames & Hudson.

Agenda (1977). Special Issue on Myth. Vol. 15, nos 2–3.

Althusser, Louis (1969). *For Marx*. Brewster, Ben (trans.). London: Allen Lane.

Althusser, Louis (1971). *Lenin and Philosophy and Other Essays*. Brewster, Ben (trans.). London: New Left Books.

Althusser, Louis & Balibar, Etienne (1977). *Reading 'Capital'*. London: New Left Books.

Anderson, Perry (1969). Components of the national culture. In Cockburn, Alexander & Blackburn, Robin (eds), *Student Power: Problems, Diagnosis, Action*. Harmondsworth: Penguin.

Atwood, Margaret (1983). *Murder in the Dark: Short Fictions, and Prose Poems*. Toronto: Coach House Press.

Austin, John (1962). *How to do Things with Words*. Oxford: Clarendon Press.

Bakhtin, M. M. (1968). *Rabelais and his World*. Iswolsky, Helene (trans.). London: MIT Press.

Bakhtin, M. M. (1981). *The Dialogic Imagination. Four Essays*. Holquist, Michael (ed.), Emerson, Caryl & Holquist, Michael (trans.). Austin: University of Texas Press.

Bakhtin, M. M. (1984). *Problems of Dostoevsky's Poetics*. Manchester: Manchester University Press.

Bakhtin, M. M. (1986). *Speech Genres and other Late Essays*. McGee, Vern W. (trans.). Austin: University of Texas Press.

Bal, Mieke (1985). *Narratology: Introduction to the Theory of Narrative*. Van Boheemen, Christine, (trans.). London: University of Toronto Press.

Balibar, Etienne & Macherey, Pierre (1978). Literature as an ideological form. McLeod, Ian, Whitehead, John & Wordsworth, Ann (trans.). *Oxford Literary Review* 3(1), 4–12.

Banfield, Ann (1982). *Unspeakable Sentences: Narration and Representation in the Language of Fiction*. London: Routledge.

Banfield, Ann (1985). Écriture, narration and the grammar of French. In Hawthorn, Jeremy (ed.), *Narrative: from Malory to Motion Pictures*. London: Arnold, 1–22.

Barrett, Michèle (1989). Some different meanings of the concept of 'difference': feminist theory and the concept of ideology. In Meese, Elizabeth & Parker, Alice (eds), *The Difference Within: Feminism and Critical Theory*. Amsterdam: John Benjamins, 37–48.

Barrett, Michèle (1991). *The Politics of Truth: From Marx to Foucault*. Oxford: Polity Press.

Barthes, Roland (1964). *On Racine*. Howard, Richard (trans.). (First published in French, 1963.) New York: Hill & Wang.

Barthes, Roland (1967a). *Elements of Semiology*. Lavers, Annette & Smith, Colin (trans.). (First published in French, 1964.) London: Cape.

Barthes, Roland (1967b). *Writing Degree Zero*. Lavers, Annette & Smith, Colin (trans.). (First published in French, 1953.) London: Cape.

Barthes, Roland (1973). *Mythologies*. Lavers, Annette (ed. & trans.). (First published in French, 1972.) Frogmore: Granada.

Barthes, Roland (1975). An introduction to the structural analysis of narrative. *New Literary History*, 6(2), 137–72.

Barthes, Roland (1976). *The Pleasure of the Text*. Miller, Richard (trans.). (First published in French, 1975.) London: Cape.

Barthes, Roland (1977). The death of the author. In Heath, Stephen (ed. & trans.), *Image-Music-Text*. London: Fontana, 142–8.

Barthes, Roland (1981a). Textual analysis of Poe's 'Valdemar'. Bennington, Geoff (trans.). In Young, Robert (ed.), *Untying the Text: A Post-structuralist Reader*. London: Routledge, 133–61.

Barthes, Roland (1981b). Theory of the text. In Young, Robert (ed.), McLeod, Ian (trans.), *Untying the Text: A Post-structuralist Reader*. London: Routledge, 31–47.

Barthes, Roland (1990). *S/Z*. Miller, Richard (trans.). (First published in French, 1973.) Oxford: Blackwell.

Belsey, Catherine (1980). *Critical Practice*. London: Methuen.

Benjamin, Walter (1973). *Illuminations*. Arendt, Hannah (ed.), Zohn, Harry (trans.). (First published in German, 1955; English trans. first published 1968.) London: Collins/Fontana.

Benvenuto, Bice & Kennedy, Roger (1986). *The Works of Jacques Lacan*. London: Free Association Books.

Blanchot, Maurice (1981). The narrative voice (the 'he', the neuter). In Davis, Lydia (trans.), *The Gaze of Orpheus and Other Literary Essays*. New York: Station Hill Press.

Bloch, Ernst, Lukács, Georg, Brecht, Bertolt, Benjamin, Walter & Adorno, Theodor (1977). *Aesthetics and Politics*. London: New Left Books.

Bloom, Harold (1973). *The Anxiety of Influence: A Theory of Poetry*. New York: Oxford University Press.

Bloom, Harold (1982). *Agon: Towards A Theory of Revisionism.* Oxford: Oxford University Press.

Bodkin, Maud (1934). *Archetypal Patterns in Poetry.* London: Oxford University Press.

Booth, Wayne C. (1961). *Rhetoric of Fiction.* London: University of Chicago Press.

Booth, Wayne C. (1988). *The Company We Keep: An Ethics of Fiction.* London: University of California Press.

Bowlt, John (1972). Introduction to special issue on *Russian Formalism. 20th Century Studies*, 7/8, December 1972.

Bremond, Claude (1966). La logique des possibles narratifs. *Communications*, 8, 60–76.

Bremond, Claude (1973). *Logique du Récit.* Paris: Seuil.

Brooke-Rose, Christine (1981). *A Rhetoric of the Unreal: Studies in Narrative and Structure, Especially of the Fantastic.* Cambridge: Cambridge University Press.

Brooks, Cleanth (1946). Empson's criticism. In Quinn, Kerker & Shattuck, Charles (eds), *Accent Anthology.* (First published, 1944.) New York: Harcourt Brace.

Brooks, Cleanth (1983). In search of the New Criticism. *The American Scholar*, 53(1), Winter 1983/4, 41–53.

Brooks, Cleanth & Warren, Robert Penn (1958). *Understanding Poetry.* (Rev. edn; first published, 1938.) New York: Henry Holt.

Brousse, Marie-Hélène (1987). Des fantasmes au fantasme. In Miller, Gérard (ed.), *Lacan.* Paris: Bordas, 107–22.

Brown, Pamela & Levinson, Stephen (1978). Universals in language usage: politeness phenomena. In Goody, Esther N., (ed.), *Questions of Politeness: Strategies in Social Interaction.* London: Cambridge University Press.

Brown, Pamela & Levinson, Stephen (1987). *Politeness: Some Universals in Language Usage.* London: Cambridge University Press.

Burden, Robert (1991). *Heart of Darkness.* London: Macmillan.

Bürger, Peter (1984). *Theory of the Avant-garde.* Shaw, Michael (trans.). Minneapolis: University of Minnesota Press.

Butor, Michael (1964). L'usage des pronoms personnels dans le roman. In *Répertoire II.* Paris: Les Editions de Minuit.

Callinicos, Alex (1989). *Against Postmodernism.* London: Polity Press.

Cameron, Deborah (1985). *Feminism and Linguistic Theory.* London: Macmillan.

Cawelti, John 1977: Literary formulas and cultural significance. In Luedke, Luther (ed.), *The Story of . . . American Culture/Contemporary Conflicts.* Deland, FL: Everett/Edwards.

Caws, Mary Ann (1985). *Reading Frames in Modern Fiction.* Princeton: Princeton University Press.

Chodorow, Nancy J. (1989). *Feminism and Psychoanalytic Theory*. London: Yale University Press.

Cioffi, Frank (1976). Intention and interpretation in criticism. (First published, 1964.) In Newton-De Molina, David (ed.), *On Literary Intention*. Edinburgh: Edinburgh University Press, 55–73.

Cixous, Hélène (1981). The laugh of the Medusa. In Marks, Elaine & de Courtivron, Isabelle (eds). Cohen, Keith & Cohen, Paula (trans.), *New French Feminisms: An Anthology*. (First published in French, 1975. Translation is of the revised French version of 1976.) New York: Schocken Books, 245–64.

Cohan, Steven & Shires, Linda M. (1988). *Telling Stories: A Theoretical Analysis of Narrative Fiction*. London: Routledge.

Cohn, Dorrit (1978). *Transparent Minds: Narrating Modes for Presenting Consciousness in Fiction*. Princeton: Princeton University Press.

Conrad, Joseph (1986). *Lord Jim*. Hampson, Robert (ed.), Watts, Cedric (introduction & notes). Harmondsworth: Penguin.

Corner, John & Richardson, Kay (1986). Documentary meanings and the discourse of interpretation. In Corner, John (ed.), *Documentary and the Mass Media*. London: Edward Arnold, 140–60.

Cranny-Francis, Anne (1990). *Feminist Fiction: Feminist Uses of Generic Fiction*. Oxford: Polity Press.

Crapanzano, Vincent (1978). Lacan's *Ecrits*. *Canto* 2, 183–91.

Crapanzano, Vincent (1992). *Hermes' Dilemma and Hamlet's Desire: On the Epistemology of Interpretation*. London: Harvard University Press.

Culler, Jonathan (1975). *Structuralist Poetics: Structuralism, Linguistics and the Study of Literature*. London: Routledge.

Culler, Jonathan (1980). Prolegomena to a theory of reading. In Suleiman, Susan R. & Crosman, Inge (eds), *The Reader in the Text*. Guildford: Princeton University Press, 46–66.

Culler, Jonathan (1981). *The Pursuit of Signs: Semiotics, Literature, Deconstruction*. London: Routledge.

Culler, Jonathan (1988). *Framing the Sign: Criticism and its Institutions*. Oxford: Blackwell.

Curle, Richard (ed.) (1928). *Conrad to a Friend: 150 Selected Letters from Joseph Conrad to Richard Curle*. Curle, Richard (introduction & notes). London: Sampson Low, Marston.

Dallenbach, Lucien (1989). *The Mirror in the Text*. Whiteley, Jeremy & Hughes, Emma (trans.). Oxford: Polity Press.

Daly, Mary (1979). *Gyn/ecology: The Metaethics of Radical Feminism*. London: The Women's Press.

Daly, Mary (1984). *Pure Lust: Elemental Feminist Philosophy*. London: The Women's Press.

De Jong, Irene I. F. (1989). *Narrators and Focalizers: The Presentation of the Story in the Iliad*. (2nd edn.) Amsterdam: Grüner.

Deleuze, Gilles & Guattari, Félix (1983). *Anti-Oedipus: Capitalism and Schizophrenia*. (First published, 1977.) Minneapolis: University of Minnesota Press.

Deleuze, Gilles (1993). Rhizome Versus Tree. Repr. in *The Deleuze Reader*. Boundas, Constantin V. (ed.). New York: Columbia University Press.

Derrida, Jacques (1973). Différance. In Allison, David B. (trans.). *Speech and Phenomena, and other Essays on Husserl's Theory of Signs*. Evanston, IL: Northwestern University Press.

Derrida, Jacques (1975). The purveyor of truth. *Yale French Studies*, 52, 31–113.

Derrida, Jacques (1976). *Of Grammatology*. Spivak, Gayatri Chakravorty (trans.). (First published in French, 1967.) London: The Johns Hopkins University Press.

Derrida, Jacques (1978). *Writing and Difference*. Bass, Alan (trans.). London: Routledge.

Derrida, Jacques (1981a). *Dissemination*. Johnson, Barbara (trans.). London: Athlone Press.

Derrida, Jacques (1981b). *Positions*. Bass, Alan (trans.). London: Athlone Press.

Diamond, Nicola (1992). Introjection. In Wright, Elizabeth (1992), 176–8.

Doane, Mary Anne (1988). *The Desire to Desire: The Woman's Film in the 1940s*. London: Macmillan.

Doubrovsky, Serge (1973). *The New Criticism in France*. Coltman, Derek (trans.). London: University of Chicago Press.

Draper, Hal (1978). *Karl Marx's Theory of Revolution: The Politics of Social Class*. New York: Monthly Review Press.

Dupriez, Bernard (1991). *A Dictionary of Literary Devices*. Halsall, Albert W., (trans. & adapted). (First published in French, 1984.) Hemel Hempstead: Harvester Wheatsheaf.

Dutton, Richard (1992). Postscript. In Wilson, Richard & Dutton, Richard (eds), *New Historicism and Renaissance Drama*. Harlow: Longman, 219–26.

Eagleton, Terry (1970). *Exiles and Emigrés*. London: Chatto & Windus.

Eagleton, Terry (1976a). *Criticism and Ideology*. London: Verso.

Eagleton, Terry (1976b). *Marxism and Literary Criticism*. London: Methuen.

Eagleton, Terry (1983). *Literary Theory: An Introduction*. Oxford: Blackwell.

Eagleton, Terry (1990). *The Ideology of the Aesthetic*. Oxford: Blackwell

Eagleton, Terry (1991). *Ideology: An Introduction*. London: Verso.

Eco, Umberto (1972). Towards a semiotic inquiry into the television message. Splendore, Paola (trans.). (First read as a paper in Italian, 1965.) *Working Papers in Cultural Studies*, 3, 103–21.

Eco, Umberto (1981). *The Role of the Reader: Explorations in the Semiotics of Texts*. London: Hutchinson.

Eichenbaum, Boris (1965). The theory of the 'formal method'. (First published in Ukrainian, 1926; this translation from the Russian version, 1927.) In

Lemon, Lee T. & Reis, Marion J. (eds & trans.), *Russian Formalist Criticism: Four Essays*. Lincoln: University of Nebraska Press.

Èjxenbaum, Boris M. (1971a). The theory of the formal method. (First published, 1927, but see previous entry.) In Matejka, Ladislav & Pomorska, Krystyna (eds), Titunik, I. R. (trans.), *Readings in Russian Poetics: Formalist and Structuralist Views*. London: MIT Press, 3–37.

Èjxenbaum, Boris M. (1971b). Literary environment. (First published, 1929.) In Matejka, Ladislav & Pomorska, Krystyna (eds), Titunik, I. R. (trans.), *Readings in Russian Poetics: Formalist and Structuralist Views*. London: MIT Press, 56–65.

Eliot, T. S. (1920). *The Sacred Wood*. London: Methuen.

Eliot, T. S. (1922). *The Waste Land*. London: Hogarth Press.

Eliot, T. S. (1961). *On Poetry and Poets*. New York: Noonday Press.

Ellmann, Mary (1979). *Thinking about Women*. (First published, 1968.) London: Virago.

Ellmann, Maud (1981). Disremembering Dedalus: *A Portrait of the Artist as a Young Man*. In Young, Robert (ed.), *Untying the Text*. London: Routledge, 189–206.

Empson, William (1961). *Seven Types of Ambiguity*. (3rd edn; first published, 1930.) Harmondsworth: Peregrine.

Empson, William (1962). Rhythm and imagery in English poetry. *British Journal of Aesthetics*, 2(1), 36–54.

Empson, William (1965). *Milton's God*. (Rev. edn; first published 1961.) London: Chatto & Windus.

Empson, William (1984). *Using Biography*. London: Chatto & Windus/Hogarth Press.

Engels, Frederick (1964). *Dialectics of Nature*. (3rd rev. edn.) Dutt, Clemens (trans.). London: Lawrence & Wishart.

Ermarth, Elizabeth Deeds (1992). *Postmodernism and the Crisis of Representational Time*. Princeton: Princeton University Press.

Felperin, Howard (1992). *The Uses of the Canon: Elizabethan Literature and Contemporary Theory*. (Repr. with corrections of 1990 edn.) Oxford: Clarendon Press.

Ferenczi, Sandor (1909). Introjection and transference. *First Contributions to Psychoanalysis*. London: Hogarth Press, 35–93.

Fetterley, Judith (1978). *The Resisting Reader: A Feminist Approach to American Fiction*. Bloomington: Indiana University Press.

Fiedler, Leslie (1966). *Love and Death in the American Novel*. (Rev. edn; first published, 1960.) New York: Stein & Day.

Fish, Stanley E. (1972). *Self-consuming Artifacts: The Experience of Seventeenth-century Literature*. Berkeley: University of California Press.

Fish, Stanley (1980). *Is There a Text in this Class? The Authority of Interpretive Communities*. London: Harvard University Press.

Fludernik, Monika (1993). *The Fictions of Languages and the Languages of Fiction: The Linguistic Representation of Speech and Consciousness.* London: Routledge.

Forgacs, David (1986). Marxist literary theories. (First published, 1982.) In Jefferson, Ann & Robey, David (eds), *Modern Literary Theory: A Comparative Introduction.* (2nd expanded edn.) London: Batsford, 166–203.

Forster, E. M. (1927). *Aspects of the Novel.* London: Arnold.

Foucault, Michel (1972). *The Archaeology of Knowledge.* Sheridan Smith, A. M. (trans.). London: Tavistock.

Foucault, Michel (1979). *Discipline and Punish: The Birth of the Prison.* Sheridan, Alan (trans.). (First published in French, 1975.) Harmondsworth: Penguin.

Foucault, Michel (1980a). *Power/Knowledge.* Brighton: Harvester.

Foucault, Michel (1980b). What is an author? (First published in English, 1977, in Bouchard, Donald F. [ed.], *Language, Counter-memory, Practice: Selected Essays and Interviews.* New York: Cornell University Press.) In Harari, J. V. (ed.), *Textual Strategies: Perspectives in Post-structuralist Criticism.* London: Methuen, 141–60.

Foucault, Michel (1981). The order of discourse. Originally Foucault's inaugural lecture, delivered at the Collège de France 2 December 1970. In Young, Robert (ed.), McLeod, Ian (trans.), *Untying the Text.* London: Routledge, 48–78.

Fowler, Roger (1990). *The Lost Girl*: discourse and focalization. In Brown, Keith (ed.), *Rethinking Lawrence.* Milton Keynes: Open University Press, 53–66.

Fox, Ralph (1979). *The Novel and the People.* (First published, 1937.) London: Lawrence & Wishart.

Freud, Sigmund (1955). The Uncanny. In Strachey, James & Freud, Anna (eds), *The Standard Edition of the Complete Psychological Works of Sigmund Freud*, Vol. XVII, 219–52.

Freud, Sigmund (1963). Female sexuality. (First published, 1931.) Repr. in Rieff, Philip (ed.), *Sigmund Freud: Sexuality and the Psychology of Love.* New York: Macmillan, 194–211.

Freud, Sigmund (1976). *The Interpretation of Dreams.* Strachey, James (ed., assisted by Alan Tyson), Strachey, James (trans.). Harmondsworth: The Pelican Freud Library, Vol. 4.

Freud, Sigmund, & Breuer, Joseph (1974). *Studies on Hysteria.* Strachey, James & Alex (eds, assisted by Angela Richards), Strachey, James & Alex (trans.). Harmondsworth: The Pelican Freud Library, Vol. 3.

Frow, John (1986). *Marxism and Literary History.* Oxford: Blackwell.

Furbank, P. N. 1970: *Reflections on the Word 'Image'*. London: Secker & Warburg.

Gadamer, Hans-Georg (1989). *Truth and Method.* (2nd, rev. edn.). Trans. rev. by Weinsheimer, Joel & Marshall, Donald G. London: Sheed & Ward.

Gagnon, Madeleine (1980). Body I. An excerpt from 'Corps I'. (First published in French, 1977.) In Marks, Elaine & Courtivron, Isabelle de (eds), Courtivron, Isabelle de (trans.), *New French Feminisms: An Anthology*. Amherst: University of Massachusetts Press.

Galton, Ray & Simpson, Alan (1987). *The Best of Hancock: Classics from the BBC Television Series*. (First published, 1986.) Harmondsworth: Penguin.

Garvin, Paul (ed. & trans.) (1964). *A Prague School Reader on Esthetics, Literary Structure, and Style*. Washington, DC: Georgetown University Press.

Geertz, Clifford (1973). *The Interpretation of Cultures*. New York: Basic Books.

Genette, Gérard (1979). *L'introduction à l'architexte*. Paris: Seuil. English trans. by Lewin, Jane E. (1992): *The Architext: An Introduction*. Berkeley: University of California Press.

Genette, Gérard (1980). *Narrative Discourse*. Lewin, Jane E. (trans.). Oxford: Blackwell.

Genette, Gérard (1982). *Figures of Literary Discourse*. Sheridan, Alan (trans.). Oxford: Blackwell.

Genette, Gérard (1987). *Seuils*. Paris: Editions de Seuil.

Gilbert, Sandra M. & Gubar, Susan (1988). *No Man's Land: The Place of the Woman Writer in the Twentieth Century. Volume 1: The War of the Words*. New Haven, CT: Yale University Press.

Goffman, Erving (1974). *Frame Analysis: An Essay on the Organization of Experience*. Cambridge: Harvard University Press.

Goldmann, Lucien (1969). *The Human Sciences and Philosophy*. White, Hayden V. & Anchor, Robert (trans.). London: Cape.

Graff, Gerald (1979). *Literature against Itself: Literary Ideas in Modern Society*. London: University of Chicago Press.

Gramsci, Antonio (1971). *Selections from the Prison Notebooks of Antonio Gramsci*. Hoare, Quintin & Nowell Smith, Geoffrey (eds & trans.). London: Lawrence & Wishart.

Green, Michael (1976). Cultural Studies at Birmingham University. In Craig, David & Heinemann, Margot (eds), *Experiments in English Teaching: New Work in Higher and Further Education*. London: Arnold, 140–51.

Greenblatt, Stephen J. (1988). *Shakespearian Negotiations*. Oxford: Clarendon Press.

Greenblatt, Stephen J. (1990). *Learning to Curse: Essays in Early Modern Culture*. London: Routledge.

Gregor, Ian (1970). Criticism as an individual activity: the approach through reading. In Bradbury, Malcolm & Palmer, David (eds), *Contemporary Criticism*, Stratford-upon-Avon Studies 12. London: Arnold, 195–214.

Gregory, R. L. (1970). *The Intelligent Eye*. London: Weidenfeld & Nicolson.

Gumperz, John (1982a). *Discourse Strategies*. London: Cambridge University Press.

Gumperz, John (1982b). *Language and Social Identity*. Cambridge: Cambridge University Press.

Hall, Stuart & Whannel, Paddy (1964). *The Popular Arts*. London: Hutchinson.

Hallberg, Robert von (1985). *Canons*. London: University of Chicago Press.

Hampton, Christopher (1990). *The Ideology of the Text*. Milton Keynes: Open University Press.

Harari, Josué V. (ed.) (1980). *Textual Strategies: Perspectives in Post-structuralist Criticism*. London: Methuen.

Harland, Richard (1987). *Superstructuralism: The Philosophy of Structuralism and Post-structuralism*. London: Methuen.

Hartman, Geoffrey (1970). *Beyond Formalism*. New Haven, CT: Yale University Press.

Harvey, David (1989). *The Condition of Postmodernity*. Oxford: Blackwell.

Hassan, Ihab (1985). The culture of postmodernism. *Theory Culture and Society*, 2(3), 119–31.

Havránek, Bohuslav (1964). The functional differentiation of the standard language. (First published in Czech, 1932.) In Garvin, Paul L. (1964).

Hawthorn, Jeremy (1987). *Unlocking the Text: Fundamental Issues in Literary Theory*. London: Arnold.

Hazlitt, William (n.d.). *Table Talk or Original Essays*. London: Everyman Library/J. M. Dent.

Henricksen, Bruce (1992). *Nomadic Voices: Conrad and the Subject of Narrative*. Urbana & Chicago: University of Illinois Press.

Hewitt, Douglas (1972). *The Approach to Fiction: Good and Bad Readings of Novels*. London: Longman.

Hirsch, E. D. (1967). *Validity in Interpretation*. London: Yale University Press.

Hirsch, E. D. (1976). *The Aims of Interpretation*. London: University of Chicago Press.

Hoggart, Richard (1957). *The Uses of Literacy. Aspects of Working-class Life, with Special Reference to Publications and Entertainments*. London: Chatto & Windus.

Hoggart, Richard (1970). Contemporary cultural studies: an approach to the study of literature and society. In Bradbury, Malcolm & Palmer, David (eds), *Contemporary Criticism*, Stratford-upon-Avon Studies 12. London: Arnold, 154–70.

Holderness, Graham (1982). *D. H. Lawrence: History, Ideology and Fiction*. Dublin and London: Gill & Macmillan.

Holderness, Graham (1991). Production, reproduction, performance: Marxism, history, theatre. In Barker, Francis, Hulme, Peter, & Iversen, Margaret (eds), *Uses of History: Marxism, Postmodernism and the Renaissance*. Manchester: Manchester University Press, 153–78.

Holderness, Graham (1992). *Shakespeare Recycled: The Making of Historical Drama*. Hemel Hempstead: Harvester Wheatsheaf.

Holland, Norman N. (1975). *5 Readers Reading*. London: Yale University Press.

Hume, Kathryn (1984). *Fantasy and Mimesis: Responses to Reality in Western Literature*. London: Methuen.

Humm, Maggie (1989). *The Dictionary of Feminist Theory*. London: Harvester Wheatsheaf.

Hunt, Alan (1977). Theory and politics in the identification of the working class. In Hunt, Alan (ed.), *Class and Class Structure*. London: Lawrence & Wishart.

Huyssen, Andreas (1988). *After the Great Divide: Modernism, Mass Culture and Postmodernism*. London: Macmillan.

Ingarden, Roman (1973). *The Literary Work of Art: An Investigation on the Borders of Ontology, Logic, and Theory of Literature*. (First published, 1931.) Grabowicz, George G. (trans.). Evanston, IL: Northwestern University Press.

Iser, Wolfgang (1974). *The Implied Reader: Patterns of Communication in Prose Fiction from Bunyan to Beckett*. London: The Johns Hopkins University Press.

Jacob, Carol (1978). *The Dissimulating Harmony: The Image of Interpretation in Nietzsche, Rilke, Artaud and Benjamin*. de Man, Paul (introduction). Baltimore: The Johns Hopkins University Press.

Jakobson, Roman (1960). Closing statement: linguistics and poetics. In Sebeok, Thomas A. (ed.), *Style in Language*. Cambridge, MA: The Technology Press/MIT & New York: John Wiley.

Jakobson, Roman (1971). On realism in art. (First published in Czech, 1921.) In Matejka, Ladislav & Pomorska, Krystyna (eds), Magassy, Karol (trans.), *Readings in Russian Poetics: Formalist and Structuralist Views*. London: MIT Press, 38–46.

Jakobson, Roman (1971). The dominant. (Delivered as a lecture in 1935.) In Matejka, Ladislav & Pomorska, Krystyna (eds), Eagle, H. (trans.), *Readings in Russian Poetics: Formalist and Structuralist Views*. London: MIT Press, 82–7.

Jakobson, Roman & Halle, Morris (1971). *Fundamentals of Language*. (2nd rev. edn.) The Hague: Mouton.

Jameson, Fredric (1981). *The Political Unconscious: Narrative as a Socially Symbolic Act*. London: Methuen.

Jauss, Hans Robert (1974). Literary history as a challenge to literary theory. In Cohen, Ralph (ed.), *New Directions in Literary History*. London: Routledge, 11–41.

Jencks, Charles (1993). Letter printed in *Times Literary Supplement* no. 4693, 12 March, 15.

Johnson, Pauline (1984). *Marxist Aesthetics: The Foundations within Everyday Life for an Enlightened Consciousness*. London: Routledge.

Kafka, Franz (1957). *The Castle*. Muir, Willa & Edwin (trans.). (First published in German, 1926). Harmondsworth: Penguin.

Keitel, Evelyne (1992). Reading as/like a woman. In Wright (1992), 371–4.

Kenner, Hugh (1978). *Joyce's Voices*. Berkeley: University of California Press.

Kermode, Frank (1989). *An Appetite for Poetry: Essays in Literary Interpretation*. London: Collins.

Kesteloot, Lilyan (ed.) (1968). *Anthologie Négro-africaine*. Collection Marabout Université. Verviers: Gérard.

Kettle, Arnold (1988). *Literature and Liberation: Selected Essays*. Martin, Graham & Nandy, Dipak (eds). Manchester: Manchester University Press.

Klein, Melanie (1988). A Contribution to the Psychogenesis of Manic-Depressive States. In Klein, Melanie, *Love, Guilt and Reparation and Other Works 1921–1945*. (First published, 1945.) London: Virago.

Knapp, Steven & Michaels, Walter Benn (1985). Against theory. In Mitchell, W. J. T. (ed.), *Against Theory: Literary Studies and the New Pragmatism*. London: University of Chicago Press.

Knights, L. C. (1937). *Drama and Society in the Age of Jonson*. London: Chatto & Windus.

Konstatinov, F. V. *et al.* (1974). *The Fundamentals of Marxist-Leninist Philosophy*. Daglish, Robert (trans.). Moscow: Progress Publishers.

Kress, Gunther & Hodge, Robert (1979). *Language as Ideology*. London: Routledge.

Kristeva, Julia (1980). *Desire in Language: A Semiotic Approach to Literature and Art*. Roudiez, Leon S. (ed.), Gora, Thomas, Jardine, Alice & Roudiez, Leon S. (trans). Oxford: Blackwell.

Kuhn, Thomas S. (1970). *The Structure of Scientific Revolutions*. (2nd edn.) London: University of Chicago Press.

Lacan, Jacques (1970). Of structure as an inmixing of an otherness prerequisite to any subject whatsover. In Macksey, Richard & Donato, Eugenio (eds), *The Languages of Criticism and the Sciences of Man*. London: The Johns Hopkins University Press, 186–95.

Lacan, Jacques (1976). Seminar on 'The Purloined Letter'. *Yale French Studies*, 48, 38–72.

Lacan, Jacques (1977). *Écrits. A Selection*. Sheridan, Alan (trans.). London: Tavistock.

Lacan, Jacques (1979). *The Four Fundamental Concepts of Psychoanalysis*. Miller, Jacques-Alain (ed.), Sheridan, Alan (trans.). (First published in French in 1973, and in English translation in 1977). Harmondsworth: Penguin Books.

Larrain, Jorge (1986). *A Reconstruction of Historical Materialism*. London: Allen & Unwin.

Lawrence, D. H. (1961). *Selected Literary Criticism*. Beal, Anthony (ed.). London: Mercury Books.

Leach, Edmund (1976). *Culture and Communication: The Logic by which Symbols are Connected.* Cambridge: Cambridge University Press.

Leavis, F. R. (1962a). *The Common Pursuit.* (First published, 1952.) Harmondsworth: Peregrine.

Leavis, F. R. (1962b). *The Great Tradition.* (First published, 1948.) Harmondsworth: Peregrine.

Leavis, F. R. (1964). *Revaluation: Tradition and Development in English Poetry.* (First published, 1936.) Harmondsworth: Peregrine.

Leavis. F. R. & Thompson, Denys (1933). *Culture and Environment: The Training of Critical Awareness.* London: Chatto & Windus.

Leavis, Q. D. (1932). *Fiction and the Reading Public.* London: Chatto & Windus.

Lehman, David (1990). Derridadaism. *Times Literary Supplement,* 18–24 May.

Leech, Geoffrey (1983). *Principles of Pragmatics.* Harlow: Longman.

Leech, Geoffrey N., & Short, Michael H. (1981). *Style in Fiction: A Linguistic Introduction to English Fictional Prose.* Harlow: Longman.

Lentricchia, Frank & McLaughlin, Thomas (eds) (1990). *Critical Terms for Literary Study.* London: University of Chicago Press.

Lessing, Doris (1973). *The Golden Notebook.* (First published, 1962.) Frogmore: Granada.

Levinson, Stephen C. (1983). *Pragmatics.* Cambridge: Cambridge University Press.

Lévi-Strauss, Claude (1972). *The Savage Mind.* London: Weidenfeld & Nicolson.

Lodge, David (1977). *The Modes of Modern Writing: Metaphor, Metonymy, and the Typology of Modern Literature.* London: Arnold.

Lodge, David (ed.) (1988). *Modern Criticism and Theory.* London: Longman.

Lotman, Yuri (1976). *Analysis of the Poetic Text.* Johnson, D. Barton (ed. & trans.). Ann Arbor, MI: Ardis.

Lovell, Terry (1980). *Pictures of Reality: Aesthetics, Politics and Pleasure.* London: British Film Institute.

Lucas, D. W. (1968). Appendix to Aristotle, *Poetics.* Oxford: Clarendon Press.

Lugowski, Clemens (1990). *Form, Individuality and the Novel: An Analysis of Narrative Structure in Early German Prose.* (First published in German, 1932.) Halliday, John Dixon (trans.). Oxford: Polity Press.

Lukács, Georg (1969). *The Historical Novel.* (First published in German, 1937.) Mitchell, Hannah & Stanley (trans). (Trans. first published, 1962.) Harmondsworth: Peregrine.

Lüthi, Max (1984). *The Fairytale as Art Form and Portrait of Man.* (First published in German, 1975.) Erickson, Jon (trans.). Bloomington: Indiana University Press.

Lyotard, Jean-François (1984). *The Postmodern Condition: A Report on Knowledge.* Bennington, Geoff & Massumi, Brian (trans.). Minneapolis: University of Minnesota Press.

Lyotard, Jean-François (1985). *Just Gaming*. Godzich, Wlad (trans.). Minneapolis: University of Minnesota Press.

MacKinnon, Catharine A. (1982). Feminism, Marxism, method, and the state: an agenda for theory. In Keohane, Nannerl O., Rosaldo, Michelle Z., & Gelpi, Barbara C. (eds), *Feminist Theory: A Critique of Ideology*. Brighton: Harvester, 1–30.

Maclean, Ian (1986). Reading and interpretation. In Jefferson, Ann & Robey, David (eds), *Modern Literary Theory: A Comparative Introduction*. (2nd edn.) London: Batsford, 122–44.

McGann, Jerome J. (1983). *A Critique of Modern Textual Criticism*. London: University of Chicago Press.

McGann, Jerome J. (1991). *The Textual Condition*. Oxford: Princeton University Press.

McHale, Brian (1987). *Postmodernist Fiction*. London: Methuen.

McLuhan, Marshall (1964). *Understanding Media: The Extensions of Man*. London: Routledge.

McQuail, Denis, Blumler, Jay G., & Brown, J. R. (1972). The television audience: a revised perspective. In McQuail, Denis (ed.), *Sociology of Mass Communications*. Harmondsworth: Penguin.

Macherey, Pierre (1978). *A Theory of Literary Production*. (First published in French, 1966.) Wall, Geoffrey (trans.). London: Routledge.

Machin, Richard & Norris, Christopher (eds) (1987). *Post-structuralist Readings of English Poetry*. Cambridge: Cambridge University Press.

Manocchio, Tony & Petitt, William (1975). *Families under Stress: A Psychological Interpretation*. London: Routledge.

Marx, Karl (1970a). *Capital. A Critique of Political Economy*. Vol. 1: *The Process of Production*. London: Lawrence & Wishart.

Marx, Karl (1970b). *Economic and Philosophical Manuscripts of 1844*. Struik, Dirk J. (ed.), Milligan, Martin (trans.). London: Lawrence & Wishart.

Marx, Karl (1971). *A Contribution to the Critique of Political Economy*. Ryazanskaya, S. W. (trans.). London: Lawrence & Wishart.

Marx, Karl & Engels, Frederick (1962). *Selected Works*. 2 vols. London: Lawrence & Wishart.

Marx, Karl & Engels, Frederick (1970). *The German Ideology*. Part 1, with selections from parts 2 & 3. Arthur, C. J. (ed.). London: Lawrence & Wishart.

Masterman, Len (ed.) (1984). *Television Mythologies: Stars, Shows and Signs*. London: Comedia.

Maturana, Humberto & Varela, Francisco (1980). *Autopoiesis and Cognition: The Realization of the Living*. Dordrecht: D. Reidel.

Medvedev, P. N./Bakhtin M. M. (1978). *The Formal Method in Literary Scholarship*. (First published in Russian, 1928.) Wehrle, Albert J. (trans.). London: The Johns Hopkins University Press.

Millard, Elaine (1989). French feminisms. In Mills, Sara, Pearce, Lynne, Spaull, Sue, & Millard, Elaine, *Feminist Reading*. Hemel Hempstead: Harvester, 153–85.

Miller, J. Hillis (1982). *Fiction and Repetition: Seven English Novels*. Oxford: Blackwell.

Miller, J. Hillis (1991). *Theory Then and Now*. Hemel Hempstead: Harvester Wheatsheaf.

Millett, Kate (1970). *Sexual Politics*. Garden City, NY: Doubleday.

Mistacco, Vicki (1980). The theory and practice of reading nouveaux romans: Robbe-Grillet's *Topologie d'une Cité Fantôme*. In Suleiman, Susan R. & Crosman, Inge (eds), *The Reader in the Text*. Guildford: Princeton University Press, 371–400.

Mitchell, Juliet (1974). *Psychoanalysis and Feminism*. London: Allen Lane.

Moi, Toril (1986). Feminist literary criticism. In Jefferson, Ann & Robey, David (eds), *Modern Literary Theory: A Comparative Introduction*. (2nd edn.) London: Batsford, 204–21.

Montrose, Louis (1989). The poetics and politics of culture. In Veeser, H. Aram (ed.), *The New Historicism*. London: Routledge.

Mukařovský, Jan (1964). Standard language and poetic language *and* The esthetics of language. (First published in Czech, 1932.) In Garvin, Paul L. (1964).

Mulhern, Francis (ed.) (1992). *Contemporary Marxist Literary Criticism*. Harlow: Longman.

Mulvey, Laura (1985). Visual pleasure and narrative cinema. In Mast, G. & Cohen C. (eds), *Film Theory and Criticism*. New York: Oxford University Press.

Nadelson, Regina (1987). Eating out with Atwood. Interview with Margaret Atwood. *The Guardian*, 18 May.

Nead, Lynda (1988). *Myths of Sexuality: Representations of Women in Victorian Britain*. Oxford: Blackwell.

Norrman, Ralf (1982). *The Insecure World of Henry James's Fiction: Intensity and Ambiguity*. London: Macmillan.

Norrman, Ralf (1985). *Samuel Butler and the Meaning of Chiasmus*. London: Macmillan.

Nuttall, A. D. (1983). *A New Mimesis: Shakespeare and the Representation of Reality*. London: Methuen.

Olsen, Stein Haugom (1978). *The Structure of Literary Understanding*. Cambridge: Cambridge University Press.

Olsen, Stein Haugom, (1987). *The End of Literary Theory*. Cambridge: Cambridge University Press.

Ong, Walter J. (1982). *Orality and Literacy: The Technologizing of the Word*. London: Methuen.

O'Toole, L. M. & Shukman, Ann (1977). A contextual glossary of formalist terminology. *Russian Poetics in Translation*, 4, 13–48.

Palmer, Paulina (1987). From 'coded mannequin' to bird woman: Angela Carter's magic flight. In Roe, Sue (ed.), *Women Reading Women's Writing*. Brighton: Harvester, 179–205.

Palmer, Richard E. (1969). *Hermeneutics: Interpretation Theory in Schleiermacher, Dilthey, Heidegger, and Gadamer*. Evanston, IL: Northwestern University Press.

Pascal, Roy (1977). *The Dual Voice: Free Indirect Speech and its Functioning in the Nineteenth-century European Novel*. Manchester: Manchester University Press.

Pavel, Thomas G. (1986). *Fictional Worlds*. London: Harvard University Press.

Pettersson, Anders (1990). *A Theory of Literary Discourse*. Lund: Lund University Press.

Picard, Raymond (1969). *New Criticism or New Fraud?* Towne, Frank (trans.). (First published in French, 1965.) Seattle: Washington State University Press.

Plekhanov, G. V. (1974). *Art and Social Life*. (Repr. of 1957 edn; first published in Russian, 1912.) Fineberg, A. (trans.). Moscow: Progress Publishers.

Plimpton, George (ed.) (1989). *Women Writers at Work: The 'Paris Review' Interviews*. Harmondsworth: Penguin.

Poggioli, Renato (1971). *The Theory of the Avant-Garde*. Fitzgerald, Gerald (trans.). New York: Harper & Row.

Pratt, Annis (1982). *Archetypal Patterns in Women's Fiction*. Brighton: Harvester.

Pratt, Mary Louise (1977). *Toward a Speech Act Theory of Literary Discourse*. Bloomington: Indiana University Press.

Prince, Gerald (1988). *A Dictionary of Narratology*. Aldershot: Scolar Press.

Propp, Vladimir (1968). *Morphology of the Folktale*. Wagner, Louis A. (ed.), Scott, Laurence (trans.). Austin: University of Texas Press.

Pynchon, Thomas (1973). *Gravity's Rainbow*. New York: Viking.

Ragland-Sullivan, Ellie (1992a). Death drive (Lacan). In Wright (1992), 57–9.

Ragland-Sullivan, Ellie (1992b). The Imaginary. In Wright (1992), 173–6.

Rayan, Krishna (1987). *Text and Sub-text: Suggestion in Literature*. London: Arnold.

Register, Cheri (1975). American feminist literary criticism: a bibliographical introduction. In Donovan, Josephine (ed.), *Feminist Literary Criticism: Explorations in Theory*. Lexington: University Press of Kentucky.

Rich, Adrienne (1976). The kingdom of the fathers. *Partisan Review*, 43(1), 17–37.

Richards, I. A. (1924). *Principles of Literary Criticism*. London: Routledge & Kegan Paul.

Richards, I. A. (1964). *Practical Criticism: A Study of Literary Judgment*. (First published, 1929.) London: Routledge.

Rickword, Edgell (1978). *Literature in Society: Essays and Opinions (II), 1931–1978*. Manchester: Carcanet.

Riffaterre, Michael (1978). *Semiotics of Poetry*. London: Methuen.

Riffaterre, Michael (1981). Interpretation and descriptive poetry: a reading of Wordsworth's 'Yew Trees'. In Young, Robert (ed.), (1981).

Rimmon-Kenan, Shlomith (1983). *Narrative Fiction: Contemporary Poetics*. London: Methuen.

Roberts, Andrew Michael (1993). Introduction to *Conrad and Gender* edition of *The Conradian*, 17(2), Spring, v–xi.

Rock, Irvin (1983). *The Logic of Perception*. London: MIT Press.

Rodway, Allan (1970). Generic criticism: the approach through type, mode and kind. In Bradbury, Malcolm & Palmer David (eds), *Contemporary Criticism*. Stratford-upon-Avon Studies 12. London: Edward Arnold, 82–105.

Rorty, Richard (1982). *Consequences of Pragmatism: Essays 1972–1980*. Brighton: Harvester.

Rosenblatt, Louise M. (1978). *The Reader, The Text, The Poem: The Transactional Theory of the Literary Work*. London: Southern Illinois University Press.

Ruthven, K. K. (1984). *Feminist Literary Studies: An Introduction*. Cambridge: Cambridge University Press.

Ryle, Gilbert (1971). The thinking of thoughts. In *Collected Papers*. Vol. II, *Collected Essays 1929–68*. London: Hutchinson, 480–96.

Said, Edward (1979). *Orientalism*. (First published, 1978.) New York: Vintage Books.

Said, Edward (1984). *The World, The Text, and The Critic*. London: Faber.

Salusinszky, Imre (1987). *Criticism in Society*. London: Methuen

Sartre, Jean-Paul (1973). *Existential Psychoanalysis*. Barnes, Hazel E. (trans.). Chicago: Henry Regnery.

Saussure, Ferdinand de (1974). *Course in General Linguistics*. Bally, Charles & Sechehaye, Albert (eds). Buskin, Wade (trans.). (Rev. edn.) London: Peter Owen.

Schank, Roger C. & Abelson, Robert P. (1977). *Scripts, Plans, Goals and Understanding: An Inquiry into Human Knowledge*. The Artificial Intelligence Series. Hillsdale, NJ: Lawrence Erlbaum Associates.

Scholes, Robert (1982). *Semiotics and Interpretation*. London: Yale University Press.

Scholes, Robert (1985). *Textual Power: Theory and the Teaching of English*. New Haven, CT: Yale University Press.

Scholes, Robert & Kellogg, Robert (1966). *The Nature of Narrative*. London: Oxford University Press.

Scott, William T. (1990). *The Possibility of Communication*. Berlin: Mouton de Gruyter.

Searle, John R. (1969). *Speech Acts: An Essay in the Philosophy of Language*. London: Cambridge University Press.

Searle, John R. (1976). A classification of illocutionary acts. *Language in Society*, 5, 1-23. (First presented as a lecture, 1971.)

Sedgwick, Eve Kosofsky (1993). *Between Men: English Literature and Male Homosocial Desire*. Reprinted with a new preface by the author. (First published, 1985.) Chichester: Columbia University Press.

Segal, Lynne (1987). *Is the Future Female? Troubled Thoughts on Contemporary Feminism*. London: Virago.

Sell, Roger (ed.) (1991). *Literary Pragmatics*. London: Routledge. (Contains Sell's own essay, 'The politeness of literary texts', 208–24.)

Seung, T. K. (1982). *Semiotics and Thematics in Hermeneutics*. New York: Columbia University Press.

Shannon, Claude Elwood (1949). *The Mathematical Theory of Communication*. Urbana: University of Illinois Press.

Sharpe, R. A. (1984). The private reader and the listening public. In Hawthorn, Jeremy (ed.), *Criticism and Critical Theory*. London: Arnold, 15–28.

Showalter, Elaine (1986). Feminist criticism in the wilderness. (First published 1981 in *Critical Inquiry*, 8, 179–205.) In Showalter, Elaine (ed.), *The New Feminist Criticism: Essays on Women, Literature and Theory*. London: Virago, 243–70.

Showalter, Elaine (1982). *A Literature of Their Own*. London: Virago.

Shklovsky, Victor (1965). Art as technique. (First published, 1917.) In Lemon, Lee T. & Reis, Marion J. (eds & trans.), *Russian Formalist Criticism: Four Essays*. Lincoln: University of Nebraska Press.

Sinfield, Alan (1992). *Faultlines: Cultural Materialism and the Politics of Dissident Reading*. Oxford: Clarendon Press.

Šklovskij, Viktor (1971). *The Mystery Novel: Dickens's 'Little Dorrit'*. (First published in Russian, 1925.) Carter, Guy (trans.). In Matejka, Ladislav & Pomorska, Krystyna (eds), *Readings in Russian Poetics: Formalist and Structuralist Views*. London: MIT Press, 220–26.

Smith, Barbara Herrnstein (1968). *Poetic Closure: A Study of How Poems End*. London: University of Chicago Press.

Smith, Barbara Herrnstein (1978). *On the Margins of Discourse: The Relation of Literature to Language*. London: University of Chicago Press.

Smith, Paul Julian (1992). Phallogocentrism. In Wright (1992), 316–18.

Solomon, Robert C. (1980). *History and Human Nature: A Philosophical Review of European Philosophy and Culture, 1750–1850*. Brighton: Harvester.

Soyinka, Wole (1984). The critic and society: Barthes, leftocracy and other mythologies. In Gates, Henry Louis Jr. (ed.), *Black Literature and Literary Theory*. London: Methuen.

Stierle, Karlheinz (1980). The reading of fictional texts. In Suleiman, Susan R. & Crosman, Inge (eds), Crosman, Inge & Zachrau, Thekla (trans.), *The Reader in the Text*. Guildford: Princeton University Press, 83–105.

Stubbs, Michael (1983). *Discourse Analysis: The Sociolinguistic Analysis of Natural Language*. Oxford: Blackwell.

Thomas, Brook (1991). *The New Historicism and Other Old-Fashioned Topics*. Princeton: Princeton University Press.

Tillyard, E. M. W. (1944)..*Shakespeare's History Plays*. London: Chatto & Windus.

Todorov, Tzvetan (1969). *Grammaire du Décaméron*. The Hague: Mouton.

Todorov, Tzvetan (1981). *Introduction to Poetics*. Howard, Richard (trans.). Brighton: Harvester.

Todorov, Tzvetan (1984). *Mikhail Bakhtin: The Dialogical Principle*. Minneapolis: University of Minnesota Press.

Tomashevsky, Boris (1965). Thematics. (First published in Russian, 1925.) In Lemon, Lee T. & Reis, Marion J. (eds & trans.), *Russian Formalist Criticism: Four Essays*. Lincoln: University of Nebraska Press.

Toolan, Michael J. (1988). *Narrative: A Critical Linguistic Introduction*. London: Routledge.

Trilling, Lionel (1966). *Beyond Culture*. London: Secker & Warburg.

Tynjanov, J. (1971). Rhythm as the constructive factor of verse. (First published in Russian, 1924.) In Matejka, Ladislav & Pomorska, Krystyna (eds), Suino, M. E. (trans.), *Readings in Russian Poetics: Formalist and Structuralist Views*. London: MIT Press, 126–35.

Tynyanov, Yu. *et al.* (1977). Formalist theory. O'Toole, L. M. & Shukman, Ann (trans.). In *Russian Poetics in Translation*, 4.

Vodička, Felix (1964). The history of the echo of literary works. (First published in Czech, 1942.) In Garvin, Paul L. (1964).

Vološinov, V. N. (1976). *Freudianism: A Marxist Critique*. Titunik, I. R. (trans.), Titunik, I. R. & Bruss, Neal H. (eds). New York: Academic Press.

Vološinov, V. N. (1986). *Marxism and the Philosophy of Language*. Matejka, Ladislav & Titunik, I. R. (trans.). London: Harvard University Press.

Vygotsky, V. S. (1986). *Thought and Language*. Kozulin, Alex (rev. & ed. version of the previous trans. by Hanfmann, Eugenia & Vakar, Gertrude, 1962). (First published in Russian, 1934.) Cambridge, MA: MIT Press.

Wain, John (ed.) (1961). *Interpretations: Essays on Twelve English Poems*. (First published, 1955.) London: Routledge.

Wales, Katie (1989). *A Dictionary of Stylistics*. Harlow: Longman.

Watson, George (1962). *The Literary Critics: A Study of English Descriptive Criticism*. Harmondsworth: Penguin.

Watt, Ian (1980). *Conrad in the Nineteenth Century*. London: Chatto & Windus.

Watzlawick, Paul, Beavin, Janet Helmick, & Jackson, Don D. (1968). *Pragmatics of Human Communication: A Study of Interactional Patterns, Pathologies, and Paradoxes.* London: Faber.

Webster, Roger (1990). *Studying Literary Theory.* London: Arnold.

Weinberg, George (1972). *Society and the Healthy Homosexual.* New York: St Martin's Press.

West, Alick (1937). *Crisis and Criticism.* London: Lawrence & Wishart.

Weston, Jessie L. (1993). *From Ritual to Romance.* (First published, 1920.) Princeton: Princeton University Press.

Wilden, Anthony (1968). *Language of the Self.* Baltimore: The Johns Hopkins University Press.

Wilden, Anthony (1972). *System and Structure: Essays in Communication and Exchange.* London: Tavistock.

Willett, John (ed. & trans.) (1964). *Brecht on Theatre: The Development of an Aesthetic.* London: Eyre Methuen.

Williams, Raymond (1958). *Culture and Society 1780–1950.* London: Chatto & Windus.

Williams, Raymond (1976). *Keywords.* Glasgow: Fontana.

Williams, Raymond (1977). *Marxism and Literature.* Oxford: Oxford University Press.

Wimsatt, W. K. (1970). *The Verbal Icon: Studies in the Meaning of Poetry.* (First published, 1954.) London: Methuen.

Wimsatt, W. K. & Beardsley, Monroe (1949). The affective fallacy. In Wimsatt, W. K. (1970).

Wimsatt, W. K. & Beardsley, Monroe (1946). The intentional fallacy. In Wimsatt, W. K. (1970).

Woolf, Virginia (1928). *Orlando.* London: Hogarth Press.

Woolf, Virginia (1929). *A Room of One's Own.* London: The Hogarth Press.

Woolf, Virginia (1966a). Professions for women. In *Collected Essays,* Vol. 2. London: Hogarth.

Woolf, Virginia (1966b). Women and fiction. In *Collected Essays,* Vol. 2. London: Hogarth.

Woolf, Virginia (1967). De Quincey's autobiography. In *Collected Essays,* Vol. 4. London: Hogarth.

Woolf, Virginia (1977). *Three Guineas.* (First published, 1938.) Harmondsworth: Penguin.

Wright, Elizabeth (ed.) (1992). *Feminism and Psychoanalysis: A Critical Dictionary.* Oxford: Blackwell.

Wright, Iain (1984). History, hermeneutics, deconstruction. In Hawthorn, Jeremy (ed.), *Criticism and Critical Theory.* London: Arnold, 83–96.

Yanarella, Ernest J. & Sigelman, Lee (eds) (1988). *Political Mythology and Popular Fiction.* Westport, CT: Greenwood.

Young, Robert (ed.) (1981). *Untying the Text.* London: Routledge.